AUGUST CELEBRATION

A Molecule of Hope for A Changing World

LINDA GROVER

Gilbert, Hoover & Clarke
PUBLISHERS

Also by Linda Grover

The House Keepers
Looking Terrific (with Emily Cho)

AUGUST CELEBRATION

A Molecule of Hope for a Changing World

Library of Congress Catalog Number 93-80687

I. Title
ISBN 0-9639538-0-X

Second Printing
Printed on recycled paper.

Book design by John David Mann
Cover photo by Gilles Arbour and Guylaine Nadeau

This book is also available on audio cassette, recorded in unabridged form by Emmy Award-winning actress Susan Clark.

Gilbert, Hoover & Clarke
PUBLISHERS
2533 Carson St. #1458
Carson City, Nevada 89706
Order Number: 1-800-900-6202

This book is dedicated to Daryl and Marta Kollman,
who, just when I was losing hope,
provided a small blue-green
model for life and love
in the 21st century.

❦

Acknowledgements

To my assistant, Vida Gabriele, whose intelligence, perception, persistence, and devotion to the project have earned her unreserved accolades. To John David Mann for his design talents and copy-editing skills, his wicked humor and infectious enthusiasm. To the most helpful reference librarians I've encountered anywhere, those of the endangered Klamath County Library. And to all those invaluable but often unsung people like Tonya Heater and Shelly Baker who help Daryl and Marta Kollman help the world.

CONTENTS

The month of August received its name
from the Roman emperor Augustus.
An “august” person was filled with
the spirit of the Goddess.
To “augur” means to prophecy—
to see—to increase.

Introduction

Imagine with me for a moment. Suppose you're just sitting around the house one day, rubbing a lamp, and a genie appears. And suppose that genie says:

"I will grant you just one wish. You have three minutes."

"Okay," you think, "what do I need most? Let's see. Something that will pay these bills . . . help end suffering and world hunger . . . and also help me lose those ten extra pounds that have been bugging me. Something that will give strength and energy to all the people I love who are feeling tired or angry or scared. Oh yes, and while I'm at it, why not something that will make me enough money so I don't have to work unless I feel like it . . ."

So you list all those items for the genie; and then you say, *"My wish is for one magical organism that can do it all."*

Then, what if the genie laughs and says, *"But that already exists, and it has since the very beginning. You just haven't noticed . . ."*

"The universe is full of magical things patiently waiting for our wits to grow sharper." —Bertrand Russell

Part I

ABUNDANT HEALTH

CHAPTER 1

From the Fires of L.A. . . .

IT WAS THE LAST DAY of the August Celebration and we basked in the sunlight, about a thousand of us, in the graceful old-fashioned park that slopes splendidly down to the shore. There were hundreds of people around me, talking, rowing on the lake, swimming, cooking their own pizzas in a brick oven, or lying on their backs and lofting a huge inflated globe in the air. It was an almost perfect picture of that partnership world of the future I'd been writing about in a novel.

Could this possibly be for real? I wondered. Are these people onto something wonderful that the rest of us just don't know about yet? The past three days had felt so sensible, peaceful, active, healthy and happy. Could this be a microcosm of an Ecotopia yet to be? Probably not, I decided.

Having just escaped from the literary jungle of network television writing, and from the L.A. riots where I watched an affluent neighbor carry looted goods into his million-dollar home, I wasn't feeling terribly hopeful about humankind that summer of 1992.

Example. My first day on the job the previous fall, I'd walked out of my vast and quite ridiculous office (with its ostentatious black leather furniture and spectacular view of the smog) and wandered down the hall to meet some of my fellow writers. But they ducked away, avoided my eye, looked nervous.

"What's this all about?" I asked when I finally cornered one of them in the Writers' Room. He was wearing a baseball cap turned sideways, the perfect uniform for a modern conformist in writerdom.

"We have a rule," he explained. "We're not supposed to talk to each other unless we're all together." He leaned forward and whispered to me; the theory seemed to be that if no two writers had a chance to link up, nobody could cut anybody else's literary throat.

To appreciate the situation you have to understand we had fifteen staff writers; I was co-head writer. Together we were supposed to turn out five hours of life-modeling drama each week, conveying every nuance of the human spirit to a national audience, millions of whom had little else in their lives to relate to. And we had to accomplish this little trick without personal contact.

Oh, we could smile at each other, we could wave, we could even exchange cute remarks at the elevator. But outside of official meetings we couldn't exchange ideas unless we lurked behind the coffee machine, or rendezvoused in the parking lot after dark. Even phone calls were forbidden. Paranoia mandated. As the weeks went by, I wondered what I'd gotten myself into. Twice before I'd quit daytime serials, both times vowing never to go back. But then it always got to be a matter of paying the bills; I often took work I didn't like.

One day, ten hours into a grueling meeting at the big round table in the Writers' Room, we were plotting an arch-villain's temporary departure from the show with a blazing shoot-em-up for a Friday climax. Then we reached an impasse—how to create a ratings boost for the mid-break commercials. (Ratings hadn't been so hot lately). Hmmm. What to do. Silence. Then we all reached for the potato chips.

Those chips were an issue. From time to time we'd all swear off, then one day the network would have them back—

they'd just appear. And pizzas. And burritos. We'd have meetings that lasted 13 hours. After a while, locked in there, drugged by the foods, you'd begin to feel grateful they were giving you anything—the Jailer's Syndrome. Years ago, on another network, they used to lock us up in suites at the Plaza or the Beverly Hills Hotel for three or four days at a time, stifling our cries for help with tiger prawns that would choke a whale. But times were tougher now.

I tore open a new bag of chips and started in on them just as the Friday outline writer said, "I've got it!" His idea in gist was that we have the villain accidentally kill his poor mother in the course of his escape. *Gratuitous violence.* "But that doesn't do a thing for our plot," I said.

"He plugs her with a stray bullet, she gasps, slumps, bleeds, forgives him, dies. Neat, huh?" Everybody cheered, and the idea was immediately voted in. *So much for creativity.*

Accuracy soon became another sore point for me. "We don't care about medical accuracy," somebody in charge told me one day with a straight face. "So we'll get a few letters. So what?"

Don't you realize we have viewers out there suffering from the very diseases you're planning to distort? And that every time you have a character shoot a handgun on daytime TV, you're teaching little kids to do the same? I wanted to say these things loud and clear, but since I'm a certified card-carrying coward, I mostly mumbled them—just loudly enough to make myself unpopular. I couldn't wait to get away.

Fortunately, just about that time I got an offer to do my own multi-cultural daily drama for the international market about a children's hospital. Non-violent character-based stuff. Unfortunately, the job would mean spending most of our time in Miami. "Sounds great," I said.

Actually it sounded kind of awful—any metropolis big enough to have parking meters is too big for Tom and me. Especially Tom, my tall and lanky, grey-pony-tailed gentle artist partner who had spent much of his life as a fishing guide. But it was better than L.A. The old Tudor house just above Sunset Boulevard that we leased was as perfect for us as it gets in L.A., but we hadn't agreed to take it till the landlady told us about the hawks that lived nearby. Tom always saw Los Angeles as a wilderness. He'd stand out on the porch and watch the hawks, and feed doves that came by the hundreds. When he looked out at the skyline, he didn't see the buildings; he saw the hills. And the trees. He'd pick out a trail and then we'd walk it—one day all the way down through Culver City to the beach in Venice. Tom could put up with L.A., but he never felt at home there. Maybe Miami would be better.

So we arranged to sublet the house to a New York producer and his Norwegian wife. We started making our deal for the TV series, but the process dragged on; the 24-page contract kept getting longer and more complicated. Late in April, the last day of the jury's deliberations in the Rodney King beating case, we were still shuttling between our attorney in Beverly Hills and the studio's firm in Westwood. When the King verdict came in, we quit negotiating and headed for the house. Or rather, we pointed our car in that direction, then inched our way through horn-honking gridlocked panic. *Not guilty!* We couldn't believe it. We could feel the outrage about to explode. Several hours later it did.

The last two nights of our stay in L.A., we alternated between scrubbing the house for the renters (have you ever tried cleaning to Norwegian standards?) and running out to the porch to see where new fires were starting. At one point there were hundreds, spread out in a 180-degree swath from Hollywood Boulevard all the way to Marina del Rey. I got to be pretty good at spotting each tiny glow as it started. Then the action got closer—first the billboard down on Melrose on top of the record store was torched, and then the guy coming to fix the Jacuzzi had a brick thrown at him less than a mile from the house. Time to get out of town.

"Pretty soon," said Tom, "nobody will go out any more. Except in their bulletproof cars to their bulletproof offices. Everything will be deliverable. Armed pizza cars."

On the morning of May first the riots finally waned and a pall of chemical-laden smoke settled on the city, dirtying everything it touched, including every surface in the house we had just scoured. We turned our keys over to the hapless renters from New York. Then before they could change their minds, we jumped into the car, sped out over Laurel Canyon and escaped north. *Failure to adjust to an insane world does not constitute insanity,* I reassured myself as we headed up I-5.

To wait out the time till our Miami deal was settled, we took temporary refuge on the farm of some friends on the northern California coast, setting up a tent in a sunlit clearing where foxes came to play every afternoon. It was a wonderfully bucolic setting where I could pursue my favorite unpaid avocation—trying to promote my own and other writers' work about a near-future world we might actually want to live in. A world where healthy people live in harmony with the environment—where they form partnerships and link together instead of always trying to dominate one another. It comforted me to live in that fantasy in my working hours. I firmly believe that without a vision the people perish—but if you've noticed, we don't seem to have one any more. Almost nobody even talks about the future. Fathers don't put an arm around their kid like they used to in the movies, point out toward the horizon and say, "Fifty years from now, Son, out thar . . ." Perhaps we don't admit it, but most of us are too scared to even look in the direction we're headed, let alone talk to our children about it.

On that farm by the ocean we spent our days trying to sell our hosts' screenplay update of Ernest Callenbach's seventies classic novel *Ecotopia.* In the evenings we'd gather in the communal kitchen of the ramshackle 150-year-old farmhouse. There we'd eat vegetables from the garden, cooked on a hot plate, and listen while our friends traded tales of anti-establishment protest—one of the residents had the unusual habit of throwing herself in front of nuclear submarines; others were into serious lobbying in Sacramento.

Then our Miami deal fell through. The studio owners, new to show business and apparently eager to rake in the bucks, had gone out on a limb with their first feature and so didn't have enough money left to supply what I needed. Or at least that's what they said. What to do now? "Stay on," said our friends.

And so we did, for a bit—and it was great. I tremendously admire people who are willing to live with few comforts and devote every penny to a cause they believe in. But sharing a kitchen and bath with ten people on a permanent basis was just not our thing. I've never understood why people doing good in the world have to be poor.

"Maybe we ought to try Klamath again," Tom suggested after a couple of weeks. Klamath Falls was the quiet Oregon lumber town where we'd lived a particularly uneventful life for several years before venturing out for that soap opera. "Maybe we should settle down there and give up on the world."

"I don't know—I'm not so sure," I told him. "Why don't we look around?"

So we headed north, traveling through the heartland of California agribiz, searching for a spot we could call home. But all we saw was more evidence of a failed system—the gray-brown polluted skies, the stench of cattle feedlots, the indestructible square tomatoes piled hundreds high in endless trucks rolling past in both directions. And all of this high-profit tasteless plenty in a state so broke it was paying its bills by IOU. Every area we looked at was too expensive, too yuppie, too frantic.

As we drove on into Oregon, the radio blared reports of Somalia starving and Serbia self-destructing in hate. I just kept thinking, "*What is the MATTER with us all—we don't seem to be able to think straight anymore. Why not?*" I didn't know it yet, but that thought would soon become my entire focus.

A few days later, at a stoplight in the dreary west side of Eugene, Tom took matters into his own hands and just started heading east. "I thought we were going to go see that realtor," I protested.

"I just want to find a motel out of town." Tom replied, as he kept on driving. Tom doesn't say much when he's driving.

A hundred miles later his hands were still clenched onto that wheel. "I think I know where you're going," I said finally. I didn't have to say more; we both knew. Klamath Falls.

"Is that okay?"

"Fine by me."

Back in our old stomping grounds, Tom and I took the usual route for Saturday night cruising—up Klamath Avenue and down Main Street. The town still had the feel of the fifties. Downtown is rather exceptional because of the cleanliness, the plum trees that display a profusion of blossoms in the spring, and the neat old buildings that nobody's renovated to death. Not yet, anyway. What used to be *The Gun Store* with the huge stuffed grizzly bear in the window is now called *All Seasons Sports* and features day-glo yuppie wear, but the rugged image of the town has barely faded. Just off Main is *The Ladies' Community Lounge*, endowed in perpetuity as a municipal females-only haven, a relic from the wild West days when a lady wasn't safe on the street.

The latest boom was back in the forties, when there were 30 sawmills in town, and it feels as though time stopped shortly after that when most of them shut down. Klamath Falls is a place where you can still walk into almost any store, hand over a check and watch the clerk stick it in the drawer just as if it were cash. No questions asked. No photo-ID. No fingerprints, no blood samples. It was good to be back.

We stopped in at Ted Swan's one-man bakery where pastries were still 35 cents, then drove up and down the streets of the original hillside settlement looking at FOR SALE signs, till we spotted an underpriced 1917 house and cabin on half an acre, six blocks up the street from the courthouse. With a total of four bathrooms, four kitchens (from its rental days) and a view of our beloved Mountain Lakes Wilderness Area, it was a heck of a deal.

We scraped some funds together, signed a contract, then headed for L.A. in a rental truck (it was impossible to reserve one down there—too many Angelinos fleeing north) to pick up the first of two loads of furniture and rock. Tom's artistic medium at the time was obsidian, so he lugged around thousands of pounds of rare volcanic rock. It was always a lot of fun to help him move.

That night we stopped back at our friends' communal farm to spend the night and pick up belongings we'd left there. In the morning as I walked along the dewy grass path to our clearing, trying to fix in my mind the foxes and the jogs and the good times we'd had there, I saw a girl with long honey-blonde hair jumping high on a giant old trampoline that had been abandoned years before. It was missing half of its springs and I hoped she wouldn't fall through.

"You must be Rowie," I said as I approached. I'd been told that houseguests were expected in from London after we went to bed, and that in fact they, too, by some extraordinary coincidence, were headed for Klamath Falls. *Nobody ever goes to Klamath Falls*. I knew that. So I asked her, "Why Klamath?" She was still jumping.

"For the August Celebration," she replied.

"What's that?" I asked.

"You don't know about the August Celebration?" With that, she did a forward somersault that terrified me, jumped off the tramp and sat down in the grass to tell me about it.

". . . it's a food," she was saying a few minutes later. "Just a whole, complete, wild food that grows naturally in the lake." She had one of those perfect complexions that English girls used to have.

"And the company that harvests it is having a convention?" I asked.

"A *celebration.*" She smiled. "They look at things a little differently."

I reflected on that. I knew the building where Cell Tech had its headquarters. You couldn't miss it; it was the flagship building in town, at the confluence of Main and Esplanade—a wedge-shaped Egyptian edifice built in the twenties as a car dealership. These days banks of phone operators sat like oarsmen on some immense galley ship headed straight down Main. I used to look in the windows as I passed, wondering what went on in there. I'd always meant to stop in and find out, but had never gotten around to it. "You say a thousand people are planning to come?"

"At least a thousand. Maybe more. They're all distributors." *Uh-oh. Sounds like multi-level marketing to me.* I didn't know much about multi-level marketing—but it seemed just evangelical enough to put me off. Too many times I'd seen the gleam in some soon-to-be-former friend's eye as he or she headed across a crowded room with that, "Have I got an opportunity for you!" line. Still I was curious. I wanted to know more.

"So what does this stuff do for you, anyway?" I wanted to know.

"Changes your *life* is what it does," she answered instantly. "You ought to try it." To hear Rowie talk, you'd think this food was some kind of magic. She said it gave people more energy, expanded and improved their lives, and even helped some people drop their addictions—to food, alcohol, drugs, tobacco, sugar, and nasty habits like bickering and negative thinking.

Then we were called in to breakfast. Rowie introduced her business partner, and her father, who'd just produced a BBC documentary called *Greenbucks*, about 'green' (i.e., environmentally responsible) companies with big earnings. During the meal, they all kept talking about "the algae," which they pronounced in the British manner, with a hard G.

"Lots of minerals. Lots of vitt-amins," said Rowie, offering a handful of green capsules around the table.

"I'll pass," said Tom. I didn't take any, either. We knew about the algae that came every year and just took over the lake. The stuff was actually quite pretty, like cut grass suspended just below the surface, and when you windsurfed over it, it was like gliding over a green patterned linoleum. But I would never have considered eating it; that was beyond my ken.

They started talking about the lake. About what an ecological paradise it was, how it was the only place in the world where this algae would grow thick enough to harvest, and how there was enough of it to feed everybody on the planet a gram a day—that it would never run out. "If everybody ate this blue-green algae, it would change the world."

"Very interesting," I said, though the whole thing smacked to me of the kind of California crystal mumbo-jumbo I'd been assiduously avoiding for years. Funny, I thought, munching on my scrambled tofu, how perfectly sensible people come to California and immediately get involved in that nonsense. And now it looked like it was spreading over the border into Oregon as well.

On the long trip south in the truck, I found myself thinking about the algae—a high-protein, high-vitamin, high-mineral food that just grew naturally for the harvesting? But how could a food possibly stop people from bickering? Ridiculous.

"So what did you think about what they were saying?" I asked Tom.

"Linda, it's a sales pitch," he replied.

Perhaps, I thought. But at least thinking about the algae kept my mind off the city that I was now bracing myself to enter again, even if only for one day. I still remembered what it was like during the riots, caught in the middle of that panicked gridlock of humanity all trying to get home to a false safety. I remembered gangs running across the intersections. And as we had passed the Beverly Center, a local icon of over-indulgence and $200 jeans, we'd seen a group of people handcuffed to the front of the building. The accusing look on the face of a young woman as her eyes met mine had stayed with me all summer.

Now, as we returned, we passed blackened buildings along Hollywood Boulevard. Back on Marmont Lane, however, the scene was maddeningly unchanged from pre-riot days. Stretch limos and Ferraris still glided by ad nauseam, while gardeners and maids still hiked down to the bus stop—nowhere is class gap more evident than in L.A.

We cleaned out my office and Tom's studio, then carefully loaded those 3,000 pounds of razor-sharp melon-sized rocks into five-gallon buckets. With my son Jamie's help we formed a bucket brigade. Then, three abreast in the truck, which was a low-rider Ryder by this time, we pulled out of the drive, bumped over the canyon and headed north, towing our '73 CJ5 Jeep behind us.

Eighteen hours later, when we finally crossed the border into Oregon, I started to breathe easier—briefly. Then we both began noticing a faint haze around the truck. "Forest fire," said Tom, as he flipped on the radio. It was bad—after a six-year drought, thousands of acres of the forest where we had loved to backpack were going up in flames. *So even our safe haven isn't safe; this feels like L.A. all over again.*

"Well," I said, "I suppose we could always rent out those four kitchens and baths, take our packs and split." But go where? Europe? Bangkok? Every place we thought about seemed to be having problems—wars, terrorism, epidemics, drought, starvation—at that moment it felt as if the whole world was burning.

We drove into town before dusk, but it was already very dark, the air so smoky it hurt our eyes. The town was jumping, however, cash registers ringing. The algae-eaters were out in force. Parked next to the usual pick-ups with the SAVE A LOGGER, EAT AN OWL bumper stickers were lots of out-of-state cars. And every motel we stopped at was plumb out of rooms. We were about ready to bed down on the rocks in the U-Haul when we pulled in at one last place to ask. But it was no use. It too was full.

Jamie and I had just started back for the truck when we heard a familiar British voice behind us. "Linda!" It was Rowie's business partner, come to move a friend from this motel to theirs.

"Does that mean there'll be an empty room?"

"Absolutely." When we got back to the desk, he fixed it up. We unhitched the jeep from the truck and collapsed into bed.

CHAPTER 2
To the Rivers of Light

MORNING DAWNED DIM AND GREY with smoke, but at least the forest fires were coming no closer. With nothing to do till our house closing later in the day, my son Jamie and I accepted Rowie's invitation to go to the old restored 1940's movie theater downtown to check out this blue-green phenomenon which had temporarily transformed the town. There, in a packed house, the Algae Celebration began with a jolt of inspiration from John Robbins, former heir to the Baskin-Robbins fortune and author of the best-selling *Diet for a New America.* Standing on an empty stage without notes, tall, rangy and spare, he told us his story.

"I remember once, working in the advertising department," he reminisced. "We'd just come up with a slogan for the year. It was called, 'We make people happy.' That would be the central theme for all our marketing for the year. Everyone was delighted—except me. I was despondent. I went home; I felt like killing myself. My father was very upset with me. He said, 'This is a great slogan. What's the matter with you?'

"I said, 'We don't make people happy. *We sell ice cream.*' Human happiness is too important, too complex, too challenging. Too meaningful. We sell a product and when people buy it, perhaps they feel good—momentarily. But in the long run, the health implications are not so happy. My father didn't want to discuss that.

"Some years later, my uncle Bert Baskin, my dad's partner and brother-in-law, died of a heart attack. I said to my dad, 'Do you think there might be any connection between the amount of ice cream he ate and his heart attack?'

"And my dad said, 'No. His ticker just got tired and stopped.'

"And then I realized the level of denial. My father's investment at all levels in the denial process—in terms of the diet/health connection—was so great it couldn't be moved.

"So I left Baskin-Robbins. I told my father I didn't want to depend on his achievements . . . I didn't want a trust fund, I didn't want to rely on his wealth, his fortune. I moved with my wife Deo to a little island off the coast of British Columbia, and built a log cabin five miles from our nearest neighbor. We cleared the land and grew almost all our own food for eight years. Neither of us had even grown so much as a parsley plant before. Now we lived on about $500 a year, and if you don't think that's a pendulum swing from where I'd been! Up to that time—in my family—"roughing it" was when room service was late. Now we were growing our own food." *Right on*, I thought.

"There is a major shift occurring in our times," Robbins went on. "It's a shift from the old way of thinking, seeing and perceiving ourselves that looks at the natural world as having value only insofar as it can be converted into revenue, commodities, merchandise. What's emerging in us, what's breaking through the resistance in us is an understanding of our interwovenness—that looks at the natural world as having value unto itself and sees our role in *partnership* with it . . . "

"*Okay!*" I whispered to Jamie, and he grinned. Partnership. For two years I'd been writing about the partnership world envisioned in Riane Eisler's *The Chalice and the Blade*. Partnership between male and female, between humans and nature, and between nations as well . . . I loved hearing that word—and Jamie knew it well.

". . . to be nurtured by that natural world and to nurture it," Robbins went on. "See, the old paradigm teaches us to objectify everything. To use people, to use ourselves, to use the world in order to project an image of who we think we're supposed to be. The new understanding knows we are whole already, knows that we were born with a song in our heart and that we're here to sing it. Knows that we're children of God, knows that we're geniuses. Knows that the creative spark in us, the good in us, is capable of whatever befalls us, whatever comes our way.

"Now I look at the crises of our times, the urgent ecological damage, the profound alienation and disintegration in our cultural value system as calls to our humanity—as the very forces needed to activate, to bring to life our heart's purpose . . ."

These were the words I needed to hear—maybe they would give this craven coward some courage. It felt as if Robbins was speaking directly to me, but then I realized he was looking around the audience and specifically addressing the Cell Tech distributors. "I want to thank you for doing with your life something that's consistent with this deeper thought a lot of us feel," he said. "This urge to bring greater health to ourselves and to our fellow human beings—to be conduits, vehicles of greater health and greater inspiration, greater awareness for this world which is so badly in need of this blessing."

"Do you think he eats the algae?" Jamie whispered to me.

"Sounds like he does," I whispered back.

I glanced around me then, taking a better look at the people Robbins had implied were heroes. Most of the 800 or so gathered in the theater seemed to be fairly young. And thin. Mostly a Birkenstock-and-Nike crowd—I'd noticed their shoes when they came in. A pretty interesting looking bunch, actually—the kind of people I generally like to know. I hadn't forgotten that they were practitioners of multi-level marketing, but I saw no gold chains, no pinky rings, no flamboyance of any kind. These people looked like activists—but activists without a chip on their shoulder. I turned my attention back to Robbins.

"I know you've all felt at times in your life's journey like you were totally alone." The guy in John Lennon glasses sitting next to me nodded vigorously. "No one around you seemed to reflect or affirm or validate the dream you carried, the needs you had, the forces that were at work within you. There are others around who will mock you, who will put obstacles in your way, who will criticize you . . ." he went on. I could understand; it must be tough trying to sell what most people think of as pond scum.

"What you need to know," Robbins continued, "is that those people are testing you, because they're the ones who can't take the step until enough others have proven it safe. Their anxiety level is too high, their comfort zone isn't big enough to include that step. In truth, you are forging the path, blazing the trail on which they later will walk. They'll only go when it's paved, if enough people have made it that way." *So these are trail-blazers? Well, perhaps . . .*

"You're not working with these people," Robbins told the Cell Tech distributors, "you're working for them. They may not eat algae. They may not eat consciously. They may not think consciously yet. But that which we are living is the human potential, and it lives in all humans. Though it may be dormant yet, by your courage you serve to make it possible for people to do what they need to do. To not have a heart attack. To have a mind and body that are more freely accessible to their heart's purpose."

When he finished, everyone was standing up, many with tears in their eyes as they applauded. And even I walked out feeling pretty impressed.

Over the next couple of days, as the Celebration continued, I made a point of getting to know a number of Cell Tech people. One evening, attending a distributors' potluck, a former literary agent turned algae distributor asked me a question. "Would you be interested in getting together with a few of us to help Cell Tech produce a video? And maybe you could even write a book. About the algae and the ecology of the lake."

Well? Hmm. Sounded interesting. I could use the money. Plus the lake would be a major feature in our new life—it was about an eight-minute jog from the house. Only one thing stopped me. Every time I asked any of the distributors, "What is it about all of you that's different—why are you so happy? What right do you have to be so well-adjusted?" their answer was basically the same.

"It's the algae."

"You've got to eat the algae."

"You wouldn't understand unless you eat the algae."

Now that really got my goat. Here was yet another form of elitism, and I reject that. *I'm just as good as you are,* I thought. *My mind functions just as well. I like you all a lot. I'll hang out with you, I'll consult with you on a video if you want. I may even write a book. But I'm perfectly healthy already. I'm not about to eat this stuff.*

Shortly after the Celebration, Rowie invited a group of us to drive up and swim at the spectacular headwaters, where she said rivers of incredibly pure water sprang right up out of the ground and flowed into the lake. We accepted, but I was feeling a little edgy as we started out. I've always disliked climbing into the back seat of a car with people I don't know and being transported by them at high speed I know not where. It's somehow disempowering.

As we headed north along Highway 97, everybody was trying to explain to Tom and me exactly why the algae was so powerful. Pointing to the caldera of Crater Lake up ahead (which from the back seat I couldn't see), they described the immense mountain that used to rise there. "Then about 7,000 years ago," a guy in the front was saying, "there was the most incredible volcanic explosion. The entire top 5,000 feet of the mountain blew off—if you can imagine that."

"They say it was the biggest explosion ever on the North American continent," somebody else chimed in.

"Like hundreds of times the size of Mount St. Helens. In fact the Indians who witnessed the explosion have passed the story down from parent to child for 300 generations."

"You're kidding," I said. That kind of cultural continuity blew my mind. In our European-based culture we don't know beans about what happened 7,000 years ago.

"And then all this rich mineral ash, millions of tons of it, came raining down on the land for hundreds of miles around."

"They've found boulders in Idaho . . ."

"And smaller pieces of rock all the way to Calgary . . ."

"Five cubic miles of earth. If you look at a map you'll see thousands of miles of wilderness—the snowmelt from all that land drains down into the lake, carrying the ash with it. So it's like a giant nutrient funnel for the algae, which soaks up all those minerals like a living sponge."

"And the sediment from the ash—and from the endless generations of algae—has built up in the bottom of the lake, till now it's about 35 feet thick."

"Just the top inch of it would be enough to keep the algae blooming for the next 60 years. Think about it!"

I was thinking about it. "And it was Daryl and Marta Kollman who figured the whole thing out? How did they get started?" My curiosity was growing.

"Daryl was a science teacher. And he noticed the kids he was trying to teach couldn't concentrate. They were all eating junk food, and he correlated the two—how the kids performed and what they ate. Then he started looking for something to help them. He went to this big mainframe computer in New Mexico and asked it what was the most nutritious food in the world."

Someone else broke in. "And the answer came up algae!"

"So he and Marta were in the forefront of algae culture at the beginning of the whole spirulina thing in the seventies, growing it in ponds in New Mexico."

"And then they found this wild source of algae that didn't have to be fed nutrients, and was much better—only the strongest survives the winter. I understand there's quite a story behind it."

I looked out the car window at the lake, wondering what the story was. I'd met Cell Tech's owners briefly at the Celebration. Daryl Kollman had been walking through the crowd talking to people. When he stopped, a group gathered instantly. Somebody asked a question and he responded with an explanation that was clear and also entertaining. "Hope is a molecule . . ." he'd said, and the phrase stuck with me. You could tell he was a teacher. "Hope is a tangible physical presence that can be identified and tracked as it makes its way around our bodies and minds," he said. He was talking about biochemistry—relating it to how we feel and what we believe is possible.

As he delivered the impromptu lecture, there was Marta Kollman, president of Cell Tech, standing outside the group which now pressed in tightly around her husband and business partner. Classy face. Really beautiful, with blonde shoulder-length hair. What was this woman like, I wondered. Trying to extricate Daryl from the crowd, she was gentle and diplomatic, looking amused as she reminded him they had a schedule and had to move on. You could see how much he wished he had more time for the lively discussion that was building.

I'd been told that, "Daryl concentrates on the vision, Marta on the business," and from what I saw it seemed to be quite true. Then someone introduced me to Marta. When she shook my hand and said, "We'll have to get together," I knew I could count on it; her manner was direct and straightforward. Talking to her for a few minutes, I also got a sense of the strength of her optimism; in fact, both she and her husband seemed to have an enormous optimism that moved right along with them wherever they went. I wanted to get to know them better. I like optimism.

We were leaving the lake now. I looked at that huge green sea spread out behind us and thought about the thousands of years it had been growing algae. *Why only here? Why would this be the only place in the world where it flourishes like this? Probably,* I thought, *for the same reasons we'd been attracted to the area.* Three hundred days of sunshine a year, and from working with a Winter Olympics group, I knew it also had precisely the ideal training altitude for the human body—4,200 feet. So maybe the altitude was just right for the algae, too. Of course I'm no scientist.

I wondered if the native people had eaten algae; I asked, but nobody knew. The rest of the trip we talked about the Indian legend of Crater Lake, in which the Chief of the Below World falls in love with the daughter of the tribal chief and promises her eternal life if she will live with him below the mountain. She refuses, then to escape his wrath she disguises herself in men's clothing and performs the heroic feats of a young brave. Of course, she does it all with consummate courage and skill—and lands a terrific lover as well. *My kind of story.* In the course of the fiery battle, the top of Mount Mazama collapses inward, leaving a great crater which then fills with water and becomes Crater Lake, that incredibly spiritual place which sometimes gets referred to as the "eighth wonder of the world."

About 40 miles north of Klamath Falls we pulled off the highway and drove in under a mix of aspen trees, ponderosas, and lodgepole pine. We all climbed out of the car and walked through the woods to a spot where huge quantities of water *do* come bursting from the earth, pouring right out of the ground to become a 60-foot-wide river just a few feet downstream.

"Awesome," was all I could say—for it was exactly that. Tom and I had hiked around the area for several years but I'd never seen anything quite like this—water so clear that when you looked through it to the bottom, it didn't even seem to be there. Six feet of water looked like six inches. I knelt down to see how cold it was, and my fingers got numb in about a second. Surely they were kidding about the swimming.

Somebody behind me was explaining to Tom that, "Water gets denser as it gets colder. Unless it freezes, of course; then it expands. In fact, water's the only substance that expands both when it freezes and when it's heated." (I was sure Tom already knew that.) But right now, at this temperature, the water is as dense as it can get—and it always comes out of the ground like this—at exactly 37 degrees." The water did look dense; it appeared to move more like a molten fluid than an ordinary liquid, its surface tension so strong that each ripple seemed to last impossibly long. And it sparkled like diamonds, holding the light and reflecting rainbows from its surface. I could see why these waters were called "The Rivers of Light." It was not a fanciful title.

"Do you know about the 'signature' of water?" somebody asked me.

"No."

"It's how hydrologists tell what aquifer or what source a particular water's coming from—the combination of minerals and other elements it contains."

"So where does this water come from?" I asked.

"That's the thing. They can't tell—it's a complete mystery. When they analyze it as it comes out of the ground, it has absolutely no minerals—nothing in it at all."

"Which it should have picked up if it were snowmelt," somebody interjected.

"But then as it warms up and expands, it's able to absorb huge amounts of minerals from all the surrounding volcanic ash as it flows into the lake."

"So where do *you* think the water comes from?" I asked Rowie as we walked along the bank.

"I don't know. I just know it's very special. Every time before I go back to England I just have to swim here. It makes me feel so good."

She is serious about the swimming. She was also being kind of mystical about the whole thing, which put Tom off a bit. Tom isn't known for being mystical in the usual sense.

A few minutes later we reached the spot they'd chosen for our swim. Rowie and her friends leapt into the water and cavorted about with great abandon. I envied them. I admired them. I also didn't want to appear to be a complete nerd and a chicken in front of these people.

"Do you want to go in?" I asked Tom.

Sensibly, Tom declined, laid his long lean frame down on the bank, and took a nap in the shade. But I had to try—I had to make the attempt. Inch by inch, I began to wade into the water. But the sharp stones hurt my feet, and the cold burned my skin. It felt like that fjord in Norway where I once almost turned myself into an iceberg. In five minutes of self-imposed suffering, I had managed to immerse myself up to my chest—but I was still holding my arms up above the water like a cormorant drying its wings. The cold took my breath away. I couldn't do it. Simply could not—no guts.

But look where you are, I admonished myself. *How far from those fires, all that hate and desperation. Here you are in the beauty of pure wilderness and creation, with people you like, and who like you. Get into that purity. Just take the plunge.* But I couldn't. *I guess you have to be a believer,* I thought as I climbed back up on the bank.

CHAPTER 3

Depleted Foods, Depleted Lives

(or, Life in the Couch Potato Generation)

RIDING BACK IN THE CAR that day, feeling as if I'd failed some crucial test, I decided the time had come to get serious. *How many times in your life*, I asked myself, *have you come upon something that seemed too good to be true—and then for that reason alone you didn't check it out?*

Of course, people tend to be like that. Once, helping Jamie on a project he dreamed up for his sixth grade class, my secretary and I stood outside my Manhattan office at 58th and 7th trying to hand out dollar bills. No conditions. Just, "Here, I want you to have this." But it was 20 minutes before we had any takers—and that was only because we followed them down the street.

Now, having stumbled upon a tiny microcosm of success in an otherwise failing—or at least flailing—world, I decided I should try to figure out what the magic was. Was it in the lake, in the algae, in the people? Or was it, as I suspected, simply in the mindset of the group, a kind of long-term mass hysteria? A giant, permanent placebo?

That the people distributing the algae actually believed, I knew. I had seen too many distributors regularly downing capsules or concentrated liquid to doubt their conviction that the algae was doing fantastic things for them. In fact, some of them had been eating the algae religiously for almost ten years. Did it help them? Of course, even if it didn't, when you *think* you feel better, then you *do* feel better—and you accomplish more. So what's the harm? *But what if it were true? What if such an elixir actually existed in these huge and readily renewable quantities? Just think of the possibilities.*

I was talking to one of the distributors over omelettes at the Saddle Rock Cafe one day. "You say it's just a food, right?"

"That's right."

"Then how can it have such incredibly dramatic effects? And so quickly? People have told me about taking one capsule and then a few hours later having some kind of transformational experience. How come?" I asked. "Food doesn't usually do that. I mean, you eat broccoli—you don't get up from the table and go bounding off with great leaps into major achievement."

He laughed. I pressed on.

"So how could the algae affect people that way? In 25 words or less."

The distributor thought for a long moment. "Because what we need used to be in the soil, and it isn't there anymore. And the algae replaces it," he said. "I think that's 20 words."

"I'm impressed."

"We're not getting the nutrients we need in our food, so our bodies and minds don't function the way they're supposed to," he continued. "But once we're supplied with the right rebuilding materials, the body does the rest. *Topsoil.* That's the key to the whole thing. Check it out."

The next day I drove out to the college library to do just that. Timidly, for I tend to be timid in libraries, I approached the steely-eyed reference librarian and promptly stumbled over my rehearsed request for, "information relating to the *Aphanizomenon flos-aquae* in Klamath Lake." (*You* try pronouncing *a-phan-i-zom-en-on-flos-aquae* in a knowledgeable and conversant tone.)

"I see," he said with a tight smile after I got it right the second time. "Are you, ah, for it or against it?" The man was definitely of the pond scum point of view, as were many Klamath residents. And that's despite the fact that Cell Tech was bringing a lot of money to the town, and providing jobs with basically zero impact on the environment (after all, algae grows back a whole lot faster than trees, sometimes doubling in hours). At the moment the man was staring at me as though I were from California.

I wasn't aware of a litmus test for library privileges, was the brave response on my lips; but as you've probably figured out by now, I tend to keep verbal challenges to a minimum. "I haven't made up my mind yet," was all I said.

The restraint paid off; Steely-eyes was actually quite helpful. While he assembled data on algae and the ecology of Klamath Lake, I pulled some reference works on topsoil and nutrition. Then I lined them up neatly on the table for a long day's work. By now I'd begun to suspect that there might really be a book for me in this subject; I'd learned to recognize the signs in myself. Sudden neatness is one of them.

There are over 30,000 species of algae, I read, 6,000 of which are classified as blue-green. Blue-green is the oldest form of algae. It's said to have been the first organism on Earth to *photosynthesize*—to use the energy of the sun to make food for itself. I read an article that said blue-green algae is what scientists search for first when they're looking for life on other planets, because blue-green algae is what gets the whole process started. Wow.

Algae comes in two kinds, micro and macro. It ranges in size from a single cell so tiny that 30 million can grow in an ounce of soil, to giant kelp more than a hundred feet long. And it lives everywhere—in tree bark, on polar ice caps, in scalding mineral springs, and even in the bodies of some animals. It's omnipresent—this great silent force in our ecosystem that quietly processes the CO_2 we use and then gives us back pure oxygen, plus a myriad of complex biochemical nutrients.

A little further on, I read that algae today provides some 90 percent of the Earth's oxygen, while trees and all the other plants provide the other ten percent. *What?* I couldn't believe that. I'd always thought it was the trees that were supposed to save us from the greenhouse effect. But in fact, I learned, algae represents 70 percent of the biomass on the planet, and some algae can double or triple in volume several times in a single day. So why don't we pay attention to it? To me, algae had always been just something to get rid of in the swimming pool.

For lunch I went to the student union, got some peanut butter crackers out of a machine, and bought some juice. When I got back to the library (which they call a "learning resource center" these days) I started reading about the history of humans eating algae, particularly seaweeds of various kinds, and especially in Japan. Sushi. Other groups, including the Aztecs, the Mayans, and more recently, the natives of Lake Chad in Africa, have also eaten various kinds of microalgae. One of the books said they dipped it up in baskets, strained it, formed it into cakes, dried it in the sun, then cooked it in a stew with millet or other grains. Unfortunately, Lake Chad is drying up—that whole area is becoming a desert.

I learned that we also use algae as a thickener in ice cream and other foods. I ran through a catalogue search on algae—there also seemed to be quite a lot of research being done on its possible *probiotic* (i.e., life-enhancing, health-supporting) qualities—particularly in its pigments. Anti-tumor effects and all kinds of interesting things.

About four o'clock I checked out some books on topsoil and nutrition, then walked across campus past the fountain and headed for the parking lot. There was no doubt that I was interested in learning about the algae—but should I try eating it?

Well, I thought, I suppose it *would* be a way to get protein without having to rely so much on meat, and that would be good. I feel guilty about the amount of rainforest acreage, or American wilderness, that gets destroyed to raise a hamburger. I also figure that if people didn't need so much territory to graze animals, they wouldn't have so many wars over land; we'd be better able to support the global population. And I know the statistics: People who don't eat meat have a lot fewer heart attacks and strokes, less cancer, diabetes and arthritis. There's no doubt that a great many of us are quietly killing ourselves with a fork.

These are all the sensible reasons for cutting down on meat—but the main reason a vegetarian diet appeals to me is because I just plain hate looking a meat animal in the eye—especially after my experience on a pig farm in Norway, where I'd gone to write a book.

In the spring shortly after I arrived, the farmer brought in two truckloads of young pigs, one of them with eyes so intelligent—one blue and one brown—that I've never forgotten her. I used to go and talk to her sometimes.

I'd been making real progress on my historical saga until one day in the fall when the neighborhood executioner arrived, garbed in black oilcloth from head to toe. Every hour he'd lead a pig out of the barn and shoot it. Sitting at my typewriter I'd jump each time. Then he and the farmer and his boys would hoist it up on the tree outside my window, bleed it, gut it, and cut it into four quarters. And all the other pigs knew exactly what was going on.

"Did the pig mind?" my son Steven asked me once when he was three or four and manfully biting into his pork chop. I could now say with certainty that indeed the pig does mind. Reliving that memory did it for me; I decided I wanted to give algae a try.

I headed downtown to Cell Tech, parked the car, and approached the building which someone from the Metropolitan Museum in New York had dubbed "the finest example of Egyptiana west of the Mississippi." Flanked by two Egyptian gods in bas relief (sculpted in the twenties), the Cell

Tech logo dominates the helm of the building. It displays an orange sun above purple snow-capped mountains, with blue water reflecting yellow sunlight, and a sky surrounded by a circle of green. Colorful, cheerful, optimistic—a typical Marta Kollman design.

Inside the main door, graceful decorated columns rose to high ceilings hung with flags and banners. The whole effect inspired confidence in my decision. In the bow, as it were, of the main floor, packing and shipping were quietly taking place. Hundreds of color-coded packages were being loaded onto hand trucks and carted out by brown-garbed UPS men. I talked to Mary, one of the Cell Tech shipping people; she arranged for me to buy some algae and fixed me up with a whole stack of literature. "Good luck," she said. "Let us know if you have any questions."

So I'm going to become an algae-eater, I thought as I headed back to the car. Which was kind of funny in a way. Of course, the phrase had always had a different context for me. I thought of those thick-lipped fish in our aquarium methodically cleaning the glass walls that confined them, gradually moving up and down, making their world bright and clean. This was another breed altogether. Or was it?

I still had one hurdle to cross. I wasn't about to start on this stuff without Tom's participation—or at least his blessing. I didn't want him to suffer with thinking that the woman he lived with was some kind of wacko health nut. Plus the algae was not cheap—so I felt we ought to agree. It would have to be both of us, or neither.

I don't like to think I planned out the evening on my way home, but I probably did. At that point Tom and I were still camped out in our big old empty house with no furniture (our tenant was doing a mini-series and so they'd be needing most of our things for a couple more months). Which was okay. We were enjoying the wide open spaces in our new home. It gave us a chance to kick up our heels and dance—to sand the old sugar pine floors and refinish the woodwork. We were up to our shoulders in wainscoting.

That night after dinner, I spread out my books on the living room floor (again neatly) and started regaling Tom, who was scraping varnish off the doors, with my just-barely-acquired knowledge. I started with the protein argument. Tom too had been trying to cut back on meat since he'd been turned on to some EarthSave literature a few months before.

". . . in fact, the whole nutritional thing is quite something," I said. "The algae tests out for huge amounts of vitamin B12, and beta carotene for the immune system—plus all the chlorophyll you could possibly get in any food."

"What does chlorophyll do?"

"Cleans the blood, I think . . . I mean, who knows what it does? There's so much that scientists don't know yet."

Tom didn't say anything.

"And the amino acids in the protein are in almost the same exact proportion as what's ideal for people. Plus dozens of minerals and trace elements from the volcanic sediment at the bottom of the lake. And we need all that because our topsoil is so depleted."

"Hmm."

Now I hit him with some numbers. I looked down at the page. "Listen to this. This is amazing. In 1948 you could buy spinach that had 158 milligrams of iron per hundred grams. But by 1965, the maximum iron they could find had dropped to *27* milligrams!" I moved a derelict lamp we'd found in the house to get better light. "*In 1973 it was averaging 2.2! That's from a hundred and fifty!* That means you'd have to eat 75 bowls of spinach to get the same amount of iron that one bowl might have given you back in '48."

"That's when the Popeye thing was really big, right?" Tom commented as he lifted the swinging door to the kitchen from its hinges. I was surprised he knew; he wasn't even born back then (I was a teenager). He carried the door to the sawhorses and laid it down. "Maybe you *did* get a jolt from spinach back then."

"And cobalt," I continued. "Did you know we need cobalt to process B12? Only parts per trillion, but we absolutely need it. And we have to have B12 or our red blood cells get weak. But nowadays, some of the vegetables we're supposed to get it from—like lettuce and cabbage and string beans—are testing out at zero cobalt. Zero." I hoped I was making an impression. Tom poured some varnish remover into a container, dipped a brush, and spread it on the door.

I went back to my reading. "There's 60-some elements that have been found in plant tissue, and we're putting back how many—four? What's in most fertilizers?"

"Potassium, calcium, nitrogen, and phosphates." Tom's wide putty knife made a neat swath through the thick brown goo on the door, exposing the bare wood underneath. I went over to look.

"That's beautiful," I said. Every job he tackles is done with such grace and efficiency—I wish I could do things that way. I continued. "So the food grows anyway, and with all the chemicals they add, it even looks pretty. You can ship it, you can play softball with it, you can jump up and down on it and nothing will happen. But nothing will happen if you eat it either. Nothing good, that is. It's empty. Remember what a tomato used to taste like? No, you wouldn't."

"So," Tom said, ignoring that last jibe, "We'll grow our own vegetables and put organic fertilizer on them. Manure. What's wrong with that?"

"What's wrong is that the manure comes from animals who eat plants that are grown in depleted soil! So you're back where you started from." I was trying to make the point—and I hoped Tom was hearing it—*that we may be at the top of the food chain, but we're getting the short end of the stick.*

"The whole topsoil thing is crucial," I said. "When you look at it over thousands of years and see what's happened ever since agriculture began, the pattern's been the same. Move into some fertile place and farm till the soil's shot, and then either die out or move on. Or, what's probably happened most of the time is a long middle period where people don't get the minerals, vitamins and stuff they need, so they just don't function as well. They get sluggish, they fight, make illogical decisions. They don't think."

"Sounds like what's happening now," said Tom.

"Exactly. That's part of why everybody's going nuts. And the worst of it is that back then there was always some place to move on to. But now the topsoil over pretty much the whole planet is exhausted. And so our food is weak, both plant and animal products—unless it's wild. Like the algae."

Tom nodded, but I could tell he wasn't quite there. And I wasn't about to try to force the issue. We both try not to do that. Over the nine years we've been together, we've learned to live by the theory of gradual consensus. If it's at all possible, we let the issue go until we agree. This time I thought it would take a while, and I was willing to wait. Then serendipity struck.

We had set up a temporary office on cardboard cartons in what used to be a back bedroom, and had called the phone company to come and string phone and fax lines to the house and cabin. The guy who arrived a couple days later was a garrulous sort who, when he found out what I did for a living, said, "Hey, I'm a writer, too!"

"Really?" I asked.

"You bet. In fact, I had a piece in the paper yesterday."

"We have yesterday's paper," I said, heading for the other room.

"Sports section!" he called after me. I pulled it out from under the layer of papers Tom was using to smear gobs of varnish on.

"Wow. 'THIS MONSTER GOT AWAY,'" I read.

Tom looked up. "What's that?"

I handed him the article. (Before we met, Tom had spent three days a week fishing over a period of 28 years.) Tom stared at the picture of the 14-foot-long fish, which had leaped high in the air; it was entirely out of the water.

"See, I threw out this minnow," the phone guy began, "fishing the bottom of the lake—"

"Klamath Lake?" asked Tom.

"That's right. And a couple minutes later I got this bite."

"What was it rigged with?"

"Barbless treble hook."

"Uh huh."

"This monster here then proceeded to drag us and our boat around for two hours and 21 minutes before the line finally broke."

"A sturgeon?" said Tom.

"Yup. And the only way we got a picture of it was somebody happened to have one of those throw-away cameras in their bag."

"Sturgeon's not native, is it," Tom observed.

"It's not. So we checked it out. Turns out Fish and Game released 200 sturgeon 12 inches long back in '56."

"That long ago?" Tom was interested now.

The guy continued, "And they think they probably weren't able to breed, since it wasn't a natural habitat."

"So that would be one of the originals?" Tom shook his head, incredulous.

"Looks that way. Well, guess I better finish up." The phone man went back out into the yard and connected us up to the world.

That night, while I was washing our two camping plates and cups at the sink, Tom was still marveling about the size of that fish. "Almost 40 years old and still going strong."

"Obviously from eating algae," I said pointedly. Tom eyed the bottle. Then—guess what? Slam dunk! Standing in our empty kitchen, Tom opened the bottle of capsules. Took out two for himself and handed me two. Then he drew a glass of water from the sink, and we swallowed them.

Our algae adventure had begun.

CHAPTER 4

The Algae in Action

THE FUNNY THING ABOUT THE algae is that you never know what it's going to do for you, so almost every result is a surprise. The day after we ate it—nothing. But the third morning we woke up, looked at each other, and said, almost at the same time, "What a wild dream that was!"

It happened the next day, and the next, until pretty soon it was taking us ten minutes every morning just to recite our dreams to each other—good dreams, richly detailed, in Technicolor and Dolby sound.

That's about all we noticed, except I found that for the first four or five days after I started eating the algae, about two in the afternoon I'd experience an intense and almost irresistible desire to locate a horizontal surface so I could curl up and go to sleep. Twice I actually did take a nap—something I've never done unpregnant, so it kind of shocked me.

But the algae people I talked to about it seemed unsurprised. "You must be the kind of person who's usually hyper. Runs herself right into the ground."

I nodded.

"So the algae will balance you out. The initial effects usually go away after a few days."

"I hope so."

"Like people who smoke sometimes start coughing like crazy for a little while. Or if you have a lot of toxins in your system, you might break out in a rash. It's just your body getting rid of all the junk it was hanging onto till it could get the nutrients it's been waiting for."

"Well. Thanks."

"Oh, and don't eat it too late in the day because you might not sleep that night," someone else added.

Actually it was these not-so-positive reactions to the algae that caused me to start believing in it—at least they proved there was definitely something to it. Sure enough, in a few days my snooze syndrome went away and I started feeling more energy, not less. I fancied it was a steadier, more centered kind of energy, but of course that may have been my imagination. What happened to Tom was not.

After a couple of weeks my dreams had tapered off, but one night about midnight I woke up. Tom wasn't in the bed. He was over by the dresser pulling on a pair of shorts.

"Where are you going?"

"Downstairs for a little while—I've got an idea about something I want to work on." His studio's in the basement. He didn't seem to want to talk about whatever it was, so I went back to sleep.

Late the next afternoon he was still at it, and for a week I saw very little of him. He explained that he'd had a dream in which he'd seen a sculpture in deer antler, complete in every detail. He couldn't rest till he had duplicated it. I watched in amazement as the work took shape. Tom had done carvings before, but this was in a whole different class.

The finished sculpture was a montage: at the base of the antler a woman sits crosslegged weaving a basket, and a huge bird flies overhead. Higher on the antler, in time lapse, we see the same woman, her basket now completed, as she stands holding it above her head, offering its contents to smaller birds as they fly in toward it. Three separate feathers wrap themselves around the tip of the antler to complete the message.

What did the carving say? It took Tom a while to figure that out. Feminine energy and giving; that was part of it. The feathers meant a circling of time, and the basket contained the abundance of the Earth. "Could it be *algae* in the basket?" I asked him. He thought it probably was.

"The Basket Maker" sold before it was half finished. Soon Tom was having dreams about kachina dolls, and often the next morning the cobbler's elves seemed to have left a new carving on his bench. Someone bought the first ten the day they were finished, brightly colored figures in miniature—prairie falcons, sun gods, and hano clowns climbing up a ladder.

Tom is still the same person with the same talent, but he's creating more abundantly, and the work is of higher quality. He's definitely one of those who was born with a song in his heart; every day for him now is a day of creating. And for somebody who used to have to sell cars for a living, that's a blessing.

For me, the results of the algae were more subtle and slower to appear. Like my driving. Now this is something I've never shared with anyone (especially my insurance agent). Though I've never had an accident, it's been sheer dumb luck that I've avoided disaster any number of times. With my mind constantly on other things, I never could remember to look in the rear-view mirror; I'd go for miles without even thinking about it. Then one day I realized I was just doing it—naturally and automatically. The whole driving process had finally become an integrated organic experience for me. (I still haven't learned to watch the gas gauge—but I also have yet to run out of gas.)

Then another interesting algae benefit showed up that I couldn't possibly doubt. There was no way to imagine it. Ever since I was big enough to see over my grandparents' card table, it's been a family tradition to do a jigsaw puzzle over the holidays. I, however, had always been so demonstrably hopeless at the task that I'd been permanently relegated to doing the edges—mostly the corners, actually.

I just couldn't keep the shape of a piece in my mind long enough to compare it with the space where it was supposed to fit. So I was constantly trying to jam mismatched pieces into places that were embarrassingly wrong. (I'd even got into the habit of 'palming' a piece so only I could see my incompetence.)

But this last Christmas was different. As soon as Tom dumped the puzzle out on the dining room table, I found I could keep up with him piece for piece. Just click click, bing bing, all day long—and remember, he's an artist with a good spatial sense. Apparently, some synapse in my brain had finally made contact. It was enough to make my Christmas.

As the months went on, my life became richer, my behavior less compulsive. Looking around the house I realized I'd become better organized; I didn't lose as many things. My dresser drawers got neater. I was busier than ever, but there was more time to get things done. Tom and I began starting work before six in the morning (*amazing!*), and 12 hours later still had plenty of energy for hours of fun till bedtime—and beyond.

This was the life that I'd always known was out there—the life that almost nobody seems lucky enough to live. Happy. Unharried. Productive. I was approaching life differently. I was accessing the better parts of myself in a very real sense.

I also noticed that food had ceased to be as important in my life. That kind of bore out my theory that the reason so many people have trouble with their weight is that the food they eat lacks nourishment—so instinct tells them to eat more and more. Eventually they become so undernourished and overstuffed that they sit around and become couch potatoes. And it's hard to think about saving the world when it's become a chore even to get up and walk to the refrigerator.

On the other hand, if you're feeling good, those feelings tend to radiate outward. I noticed I was establishing a closer relationship with my kids, two of whom had also started eating the algae. I hoped it would do for them all that it had done for me, but it was too early to tell. I realized that I harbor some guilt about the way I fed them growing up. I never gave them sugar, but I cringe to think of those "healthy" breakfasts every morning swimming in grease.

I worried about the kids. Steven, my architect-engineer-violinist, had been struggling with chronic fatigue for several years. And Jamie, my blond harvest-diver-boat-captain-environmentalist, had worked for years to improve a reading pace that slowed his studies. Cindy, my television-actress-ski-instructor-generous-hearted beauty, was newly pregnant and wouldn't start eating it right away. I hoped eventually it would help her cope with trying to raise a family with next to no money on Maui, where life is about as expensive as it gets.

Pretty well convinced that the algae had value, I began work on the book I'd been thinking about writing on the subject. Looking through records down at Cell Tech, I found all kinds of reinforcement for my new belief that algae could have a significant effect on both performance and behavior. I pulled out the folders labeled "Athletes" and "Children" because they're some of my favorite kinds of people.

In the Athletes file I read about the first vegetarian to climb Mount Everest; he did it on Super Blue Green Algae. Here was a major league baseball player who was an algae fan. And an ultra-marathon runner who swore by the stuff. And here was Bernard Voyer, one of the skiers who traversed Ellesmere Island at the North Pole—the equivalent of crossing three ice floes each the size of Switzerland. His wife, who had planned his diet, is an expedition nutritionist who evidently knew how to pack energy into every ounce of rations.

I read the statement of Canadian Masters ski racer Bob Switzer, who not only eats the algae, but makes a bundle marketing it to other skiers. "There's a whole group of us in ski racing who are doing really well and a lot of people are looking at us. At first it was, 'That algae stuff,' but now it's,

'What *IS* that algae stuff?'" On the other hand, here was a cyclist who preferred to keep the whole thing secret. He credited the algae with cutting seconds off his previous record. "Those seconds made the difference," he said. "They put the prize money in my pocket instead of someone else's, so why should I give that advantage away?"

This material on sports performance was fascinating to me—a decidedly amateur athlete who's always trying to improve her running. I can't break nine-minute miles no matter what I do. In fact, a fellow runner once observed, "If Linda fell out of an airplane, she'd probably fall at nine minutes a mile." I read on.

"We had a race for the world stair-climbing championship for the United Way," said professional triathlete Harreson Martell (formerly known as Marco Verchere). "Two thousand athletes from around the world came to the event." I skimmed down the page. "Often when athletes get psyched up for a race or a competition, they have so much energy that it gets diffused all over the place like a shotgun," Martell said. "But with the algae, I was fine-tuned and sharpened on running up these stairs as fast as I could without any distractions; it was like a relaxed intensity—I was keyed into the moment. The gun went off and I won the race." *Wow, this is encouraging.* I kept reading. "One thing that I find the algae really helps with is that it keeps away what we call 'the bonk.' That's what endurance cyclists call it when your blood sugar drops. You have a real drop in energy. It's like hypoglycemia; you have such a craving for food; you want to eat, and eat right now. I find that when I'm eating the algae, I don't get "the bonk."

Okay. That made sense. Just good nutrition. I put the Athletes file back in the drawer, closed it, and opened the Children folder.

I looked first at two unsolicited letters from YMCA caregivers who had taken care of Cell Tech distributors' "algae kids" during the August Celebration. One of them said, "From the time I first saw the children I knew they were different. They had an unusual air about them, a peacefulness, a joy . . . These children were quick to smile, excited about the convention and proud to be algae-eaters—their joy and love of life was transparent."

Another YMCA babysitter said she was unfamiliar with Cell Tech products, ". . . so I can only comment on what I've seen with my own eyes. I don't know exactly what it is that makes your children so special, but I know that a world of children like yours would be a beautiful, beautiful place. It would be the kind of world and existence I believe God intended." Pretty strong statement!

I also read a letter that had just come in from a Nancy Dillon in Penn Valley, California. "Aside from what the algae has done for my own life, what thrills me the most is what it's done for my 11-year-old son Janaka, who'd been a junk food vegetarian for about two years. Whatever had sugar, white flour or chocolate in it, he liked. I grieved to see his small stature, the dark shadows under his eyes, lack of appetite and motivation, poor grades, cranky and bored personality. I knew he was suffering from malnutrition, but I was helpless, because he refused to eat anything wholesome; he tossed away dozens of nice lunches I packed for him, sneaking candy and cookies from schoolmates. And he was a mean devil if he didn't get his Cocoa Crispies for breakfast. But one day, after praying for an answer, I got the revelation to PAY HIM TO EAT THE ALGAE! Up till now he would have nothing to do with it, but once I flashed five bucks in front of his eyes, he began to eat it . . . reluctantly. But he did it.

"Within a week he was doing his homework without being coerced. He hasn't been late to school since, and his grades have improved. Best of all, he's reverted to his original joyful self that I knew and loved long before he was old enough to rule the roost with his terrible dietary quirks. He even seems to have grown taller and stronger within a month of eating the algae—and his face is starting to glow. He's taken to eating oatmeal for breakfast and wants an apple in his lunch. His teachers are very pleased with the change." *Smart woman*, I thought.

I also found a folder of stories about dogs, cats, race horses, even a family of pet mice, whose health improvements on the algae were dramatic, and measured by vets—no chance for a placebo effect here. I looked through more stacks of letters, articles, and clippings. It was a whole new genre of literature for me—*Algae Stories*.

I replaced the files and headed upstairs to the interview I'd scheduled with Daryl for the book. Visiting with Daryl is always fun. You get to talking, then he'll digress into some new theory that's just occurring to him, and all of a sudden you're off on an exciting tangent, trying to keep up with a mind that's like a meteor. It's a workout.

Most of Daryl's ideas come to him early in the morning. He usually arrives at his office about 7 AM to study, to write, to think, and to plan. Freed from many of the burdens of the day-to-day management of Cell Tech (which to Marta are a pleasure), Daryl has both the time and the inclination to generate a variety of original ideas and projects.

The one that interests me most is his plan for algae ponds all over the world to reverse the greenhouse effect. He says we lost our window of opportunity to accomplish that with trees back in the seventies. But as Daryl explains, algae multiplies so quickly and produces so much oxygen per square foot that ponds with a total surface area five times the size of Colorado would be enough to start to reverse our growing CO_2 problem. And after all, it was blue-green algae that rescued us from the first greenhouse effect about three and a half billion years ago.

Daryl's algae pond plan was one of the subjects he had talked about at our last meeting; I wondered what he'd be telling me about today. When I entered his office he was sitting at the huge conference table surrounded by books, drafting a speech on a yellow legal pad; he looked like a professor in a library. Fifty-three years old and balding on top (he lost the hair *long* before he started eating algae, he points out), Daryl bounds around the office in his size 14 athletic shoes with the energy of a kid. You'll almost never find him sitting at his desk—unless he's on the phone (which isn't often). If he's writing or thinking, he prefers the more informal conference table.

On bookshelves, interspersed with a variety of science and ecology books, as well as Eastern and Western philosophy works, are rows of mementos from people whose lives he's touched—there is a whole row of baby pictures. On the coffee table there's a basket of various indoor and outdoor balls that see frequent use both in the office and outside—you never know when something's going to come whizzing past you. Daryl still occasionally plays quarterback on an actual tackle football team that gets together Sunday afternoons; but since it's a Cell Tech team, Marta says they don't usually tackle the boss. Or if they do, they do it gently.

There's a giant map of the world on the wall behind his desk, and a huge mural of a forest covering the other walls. He explains that he has no view—the only windows look out onto an air shaft. But it's about the only area of the building still available for offices; the rest is fully occupied by the various departments of the burgeoning company. And Cell Tech is growing so fast that before long we might find Daryl sitting out on the sidewalk or under a tree in the parking lot (though one could imagine him working out there quite happily).

We began our conversation—as I'd hoped—by talking about hope. "Hope gives us energy—it physically gives us energy," Daryl said. "There are more than 50 chemicals in the brain that alter our behavior and mood. The emotion mediating area of the brain is enriched with a number of different types of neuropeptide receptors. When we have the right nutrition, they supply us with hope."

"So our natural state is to be hopeful and resourceful," I said.

"Of course—the trick is to extend that hope beyond a momentary flash of brightness, so it's always an available biological possibility—which it can be if we have the algae. The way I visualize it, every person has a light when they're born, and they carry that light with them throughout their life. The quality of their life, the strength of it, depends on the quality of the light. For some people, because of their circumstances, the light goes out early in life. They lose hope; they're in a state of learned hopelessness. But if their light is really working, shining, then the quality of their life is high, their sphere of influence is large and extends to a lot of different places."

Certainly Daryl's light is bright, I thought; no doubt about that. He's a man of unbridled, unabashed enthusiasm for ideas, positive thinking, algae, celebration and success. The twinkle in his eye is such a strong element in his personality you can't miss seeing it even on an audio tape. He doesn't hold much with negativity. Yes, he knows all the bad news that's out there; he's got it all catalogued and filed away in his brain, but he doesn't dwell on it, preferring instead to focus on solutions. I've noticed that if anyone on his staff is troubled, he'll stop what he's doing to talk—then usually develop some new self-help idea from the problem. The two of them get so excited talking about the new theory he's come up with that the problem then pales in comparison.

Daryl continued. "One of the things I'm trying to do with my life now is to offer other people a choice—an opportunity to increase the quality of their life; that's really all you can do. An opportunity to get out of a rut, to move from competition to cooperation to celebration—with everyone. No one should have to climb over anyone's back to advance."

"That's a theme in your life, isn't it," I ventured. "Celebration."

"It's the key to everything," he said with one of those impossibly contagious grins. "All you have to do is celebrate, and the rest will follow like links in a chain. Celebration produces joy. Joy gives birth to hope. Hope fuels creation. Creation begins with an idea. An idea initiates 'the process.' 'The process' assures the completion of the desired outcome.

Completion of the desired outcome is the cause for celebration. Celebration produces joy . . . Think about it," he said. *"I've discovered the only way to feel joy is to celebrate.* Is there any other way?"

I thought about it. "I guess you're right."

"And if you're in a state of constant and continuing celebration, don't you suppose other people will want to be around you? You can celebrate alone or with a billion people. My experience is that with more people, the joy is magnified."

As usual, the hour slipped by quickly. Daryl has a way of encapsulating his thoughts in easy-to-follow steps. Same way with his global vision. He's a visionary, but a highly pragmatic one—he's got it all down to specific things people can do. And that seems to me to be something we need right now in this virtually leaderless world. I'd been focusing in my writing on our need for a vision, but it's clear we're also going to need step-by-step directions to get there. *It's no wonder that people come thousands of miles to see Daryl, or to hear him talk*, I thought as we said goodbye.

I went downstairs to the huge, open ordering and shipping bay. As I headed out, Marta waved brightly at me from her glassed-in office on the bridge. While Daryl—unquestioned star of the Cell Tech operation—sits in his cerebral forested hideaway upstairs, Marta rubs elbows, tends to details, and thinks out the overall picture. She listens to Daryl's schemes and ideas, then examines their practical ramifications. Yet when Marta is convinced, Daryl told me, she's the one who urges that they take the giant leap, the great gamble.

Marta has the same early morning energy all day long as her husband. When she's out on the floor packing a box or fixing a copier, a stranger would never pick her out as the boss. Anything anyone else will do, she'll do, and she'll probably do it first, whether it's putting up a staff mailbox or painting names on the parking spaces. It's clear that her office associates are friends, not subordinates. In these qualities, she seems to be a model for the 21st century boss.

By now I was convinced that both Daryl and Marta were uncommon people. Not perfect—they're as fallible and human as the rest of us—but they're doing everything they can to improve things, which is all any of us can do. Not long ago I was at a Cell Tech gathering where we were all asked, "What is the greatest gift you could ever give?" Daryl said, "Hope." When Marta's turn came she said, "Just keep Cell Tech going." I genuinely appreciate people who when the lifeboat is sinking, don't petition the government, or depend on someone else to do something. They just start bailing!

Daryl's sense is that ours is the pivotal generation and that we have just a few years left in which to extend our window of opportunity for the next millennium. He sees his goal as helping to provide pockets of hope all around the world for a more peaceful, egalitarian world of the future. And by utilizing network marketing, as of this writing he's developed over 30,000 ambassadors of hope out there—so he's making some progress toward that end.

Both Daryl and Marta seem far more interested in effecting change than making money, but they're wise enough to know it's going to take a lot of growth in the company to accomplish all the change they want. The vision is big, but Daryl's five-step program is starkly practical—and of course, both the supply of algae and the market for it are infinite. So it actually could work. As the ancient Upanishad saying goes, "Abundance is scooped from abundance and abundance remains." Here's how Daryl outlines his plan:

1. Eat the algae.
2. Share the algae.
3. Share the network marketing opportunity.
4. Build teams and families.
5. Participate globally.

I pondered the first two actions, the only ones with which I was personally familiar. Well, I could believe that just *eating the algae* would help some. If you never did another thing, at least you'd probably feel better. You might not hit your kid, or fight with your spouse as much. Maybe you'd also cut down on those rib-eye steaks and lower your cholesterol. Or strengthen your immune system with all those trace elements, and so not add to our soaring medical costs later on.

To *share the algae*, of course—at least for me—was pure automatic reflex. By now there was no way I would want anyone I loved to go without it. So with Step 2, the benefits of the algae are shared by more people.

Step 3, *sharing the opportunity*, could work as well, I realized. A good illustration of that was the story of a guy I met recently, Canadian distributor Gilles Arbour.

A number of years ago, Gilles was the owner of a restaurant in Montreal. The restaurant employed a busboy named Art Robbins, and the two became friends. Time went by and Art worked his way up to the position of chef. Later both men got divorces and they shared an apartment for a bit. Eventually Art moved away and they lost touch for several years.

Then Gilles, who didn't have much money at the time, got a letter from Art, offering him a chance to get into the algae network. (For $25, I learned, anyone can start their own algae business.) Today, Gilles and his Cell Tech sponsor Art have thousands of distributors in their networks, providing a substantial income for both. (Gilles has kept that lucky letter for years.) So Step 3 passed muster as well.

That leads us to Step 4, *building teams and families*. Daryl and Marta have encouraged the development of close to 50 Distributor Empowerment Teams all across North America. Art is active in the one in Boston, Gilles in Montreal. Members of these regional groups help one another in their Cell Tech businesses, and also initiate and organize community activities, such as highway clean-up operations and the planting of trees by the hundreds of thousands.

Distributor Empowerment Teams also *participate globally* (Step 5) through Marta's "Ten Percent Program," in which ten percent of the algae harvest is set aside for humanitarian aid. In the fall of '92, when distributor Pauline Richard and other members of the Quebec team undertook to supply algae and other assistance to about 2,000 children in an impoverished area of Nicaragua, Gilles helped to organize the trip. By this time, he knew about the beautiful things the algae had done for children, his own and those of his friends, and he wanted to be part of extending those benefits to others.

Nowhere is it more clearly evident that adequate nutrition for children is crucial to the future of our world than in Nandaime, a small town in the southwest part of a country where 60 percent of the population is under 15 years of age—and more than 75 percent of the people live in abject poverty. Due to a contaminated water supply, 90 percent of the Nandaime population suffers from parasite infestation; it's the leading cause of infant deaths in that area. Gilles remembers one beautiful little girl with haunting, lifeless eyes—when he touched her hair, a handful of it came loose. She was one of many children who often had nothing to eat but tortillas and salt, and were losing consciousness in the middle of the day at school.

Cell Tech has been providing Super Blue Green Algae to children and families in southwest Nicaragua now for over a year, working closely with a non-profit group called SPIRALE, and with a legendary Catholic priest from Quebec—the people call him Santiago. Under Santiago's leadership, teenagers distribute the algae in the barrios, and it's also provided to children through the school system.

In one abysmally poor barrio school in the city of Nandaime, children have been getting the algae every day in a glass of milk. Within three months, attendance was up 15 percent. And average test results were up, too, jumping from a score of 64 percent in 1991 to 75 percent in 1992. During the same period, at a similar but slightly more affluent school, where students received the milk but not the algae, scores remained flat at 68 percent. I sent a fax to the school administrator to check it out, and she faxed me back to say it was all true.

"The school director talked to us about the 1,300 kids eating Super Blue Green Algae," says Gilles. "She described their ability to stay quiet and concentrate in class, then to play games full out and manifest more joy. She told of the mothers' joy. Then came her question. In a shaky voice she asked, 'Mr. Kollman, we are very happy with the Super Blue Green Algae, but we are concerned about its availability. Are you going to keep sending it to us?' I can still hear Daryl's 'YES,' his voice conveying unshakable commitment. A year ago in Quebec City, when a group of distributors first came to her with the request, Marta Kollman had launched the project with the same powerful 'YES.'"

In the village of El Paso, where the algae is also distributed, there had recently been a number of cases of cholera, caused by amoebas and other parasites in the water. Water generally was brought to the village center in large earthenware pots on the back of a wooden-wheeled ox-drawn cart, then carried to each house. It was a tremendously labor-intensive process. To obtain a safe drinking water supply, villagers brought rocks from miles away and screened them by hand to make cement and build filtration beds as part of a SPIRALE-sponsored water project.

Jim Carpenter, Marta's brother and Cell Tech's *ecoordinator*, accompanied Daryl to Nicaragua. He and Daryl were impressed with the design and workmanship of the new filtration system. Solar panels now provide the energy to pump the water into a tank before it's piped into a number of homes, something which Jim says is absolutely revolutionary in the area.

Jim is a strong advocate of down-to-earth appropriate technology. When he saw that in Nandaime it takes an entire day to find and gather a three-day supply of firewood for cooking, a job that by necessity is done by the smallest children, he designed the algae shipping cartons so that with the addition of foil and glass, they can be converted into solar ovens after they're unpacked.

The Nicaragua project is taking place in a country that's been devastated by a series of natural disasters and more than a decade of civil war. A number of years ago an earthquake destroyed the capital city of Managua, and it's never been rebuilt. Then a hurricane ruined croplands. Finally, over 60,000 people were killed in the country's civil war, most of them men—and the population of tiny Nicaragua is only four million. So a great many families are now without a father.

Nicaragua isn't the only place where Cell Tech has sent humanitarian aid. I watched one day as everyone on the Home Team staff wrote greetings on several large barrels of freeze-dried algae, which were then sent off to Chernobyl, where the it's being fed to children suffering radiation damage from the nuclear accident there. And other programs are in the works.

What appeals to me about this whole enterprise is that it's self-financed and self-supporting. Have you ever noticed that most people who want to save the world begin by *asking you for money?* Cell Tech generates its own income to do good works, and frankly, that's refreshing. Daryl and Marta walk their talk—and they've built a multi-million-dollar-a-month business in the bargain.

In one of my first conversations with Daryl, feeling kind of gauche and gee-whizzish, I had said to him, "I never knew algae could do so much. Or that it was originally responsible for all life on Earth. It's just amazing."

"Yes it is," he agreed. "It's pretty incredible." We sat quietly for a moment, nodding to ourselves.

Then I blurted out, "It's so . . . (pause) *convenient!*"

Daryl burst out laughing and didn't stop for quite a while. I would soon become familiar with that laugh of his—the way he always laughs when he feels there's something he's taught that has just been completely understood. “Now,” he said, “you'll see why Marta and I had to do what we did.”

“Tell me about it,” I said. I really wanted to know—to learn the secret of their success. I mean, how many of us have had a far-out dream of doing something that would benefit ourselves and a great many other people too? I know I have. But how many of us have ever seen that vision come to reality? I think it's important to know how Daryl and Marta accomplished it all.

Part II

Abundant Hope

CHAPTER 5

A Boy and a River

IT WAS A CHILDHOOD TO put in a memory book—running through his uncle's corn and wheat fields, swimming in the river from the first "Dare you!" days of spring to the quiet times of autumn when the water was warm and lazy. Hunting with his father in the woods where coyotes, foxes, possums and jackrabbits were still abundant. Summers were long and sultry, and fragrant with earth smells like vine-ripened home-grown tomatoes. But it was an era that was about to end, and Daryl Kollman watched it happen.

He was born and raised in an agricultural object lesson called Iowa. In 1939 it was possibly as fertile a land as you could find anywhere. Sixty percent of the soil was dark drift, the finest soil on earth, and most of the rain fell obediently during the growing season. The state was laced with a network of powerful rivers, and crops sprang out of the black soil, sturdy and strong. Hogs and beef cattle fattened on Iowa corn.

Daryl grew up right on the Cedar River, which flows south from its origins north of Austin, Minnesota, through miles of farmland and then the city of Waterloo in Blackhawk County, before joining the Iowa and at last the Mississippi. When he was a child, the water was still clear; by the time he was 20 it was gray from bank to bank—the river was dead, and the state's precious soil was almost lifeless as well. Being witness to this loss helped to shape Daryl Kollman's understanding.

One day in 1944 in a working-class section of Waterloo, Iowa, while his family carried belongings from their '36 Ford into the three-bedroom no-bath house on River Street that would be their new home, a small boy with straight blond hair stood on the bank looking down into the water, studying its currents. He could see sand on the bottom, and rock. Some mud, too. The shadowy forms of fish swam by, some small, some pretty big. He was utterly captured by it all.

Through the year he watched the cycles of the river—the freezing, the melting, the way it eddied and flowed. He was especially fascinated when the river flooded, brown water rich with the Iowa topsoil it was carrying to the ocean. Once he grew so intrigued with the force of the water as it rushed past the ridge at the edge of the yard, that he began digging through, carving a channel to see if it would follow. It did; in fact it became unstoppable; soon half the Cedar River seemed to have poured into the yard and had begun seeping into the house as well.

That back yard became a crucible of learning for Daryl. Sloping down from road to river, it offered a flat area to play ball in back, and a hill down to the river that in winter drew sledders like a magnet from all over the neighborhood. It was a world made for experimenting, and just about everything Daryl and his brother Vic did was an experiment; both boys were destined to be scientists.

The Kollman family was second generation German; his parents still spoke German around the house. Daryl was the youngest of four—besides his brother Vic, he had two older sisters. His father worked at the Rath meat-packing plant straight across the river on the opposite bank. The plant was eight blocks long and had almost 2,000 workers. It also had underground pipes which fed its wastes directly into the river; all the kids knew where those pipes were. But the current was strong; the blood and everything else was carried swiftly downstream. In those days you could still swim underwater and see for a long way.

On summer days Daryl and Vic would fish from the bank, catching catfish, carp, and northern pike. Then they'd take them in to their mother, who'd wash and clean them at the kitchen pump, then bread and fry them in a skillet; they ate them with tomatoes from the garden.

Despite its riverside location, the neighborhood was rough. Everyone was poor, and a lot of people drank too much. About a block away was a house where the grown kids in the family were always getting hauled off to jail—for petty crimes, for armed robbery, and then for murder. Daryl and the other kids saw the police cars and they overheard their parents talking about it at night.

Daryl's mother wanted to move up and out and on to better things, but his father had no dreams, no expectations, no belief things could ever get better. He never took a chance on anything and never spent a dime unless he had to, to the great frustration of his wife. But she made the best of it. On Sunday she scrubbed up the kids and took them to church, often picking up some of the scruffiest of their neighbors to take along. Those neighbors sometimes smelled terrible, Daryl recalls. Later he learned to respect his mother for her principles, but at the time he just wanted to get away from it all.

When he was six, Daryl escaped to the mostly middle class world of the brick schoolhouse at the top of the hill. He wanted to belong, to be accepted by the other kids, but his clothes and the neighborhood he came from set him apart.

Between first and second grade Daryl and his family did something different. At his mother's insistence they finally drove in that '36 Ford all the way out to California to see relatives. Maybe relatives is what the rest of the family saw, but what Daryl saw was the OCEAN! It was impossible, unbelievable. So huge. It was the most powerful thing he'd ever seen—or imagined. Up to that time Daryl's world had been the river; now that world had expanded exponentially.

Riding home, Daryl stored it all up—everything about the beach and the rocks and the crashing surf. The sea shells, the kelp pods drying on the sand. The smell of it. The way the sun shone through a cresting wave, and the salt that dried on his skin. The round smooth stones that he picked up and carried in his pocket. Everything.

When he got back to school in the fall he told his second grade class about it all in great and perfect detail. He *had* to, because nobody in the class had ever seen the ocean. And everyone not only listened—they wanted to know more. That day Daryl's sense of poverty vanished. He learned that the wealth of knowledge could make more of an impression than a new shirt.

Though Daryl's father was frugal with money, he was generous with love. He was athletic, and spent long hours playing with the kids and teaching them sports. Some of Daryl's best times were spent walking along the river with his dad, looking for firewood which they'd mix with coal in the pot-bellied stove that was their central heat. Or digging in the garden with him in the rich fragrant soil alive with earthworms. By the time he was nine, he was taking a man's part on the two-man saw, cutting up wood for winter.

One of the things his father honored was when he didn't show pain. If he was hurt in some way and didn't complain, he was rewarded. So Daryl got careless about being hurt, and as he grew up, he became quite reckless, agile, and swift. He was his mother's greatest trial. "Daryl, don't!" became the phrase most often heard around the house, but often as not, Daryl wouldn't be there to hear her plea. He'd learned to time it; he could be a block away and out of earshot by the time the screen door slammed and his mother got to there to call after him, "Daryl, don't!"

Daryl was learning about the rate of speed and force of a moving object, whether it was himself hurtling down sheer ice on skates, or his knife thrown thousands of times into one of the two giant cottonwoods in the back yard which absorbed much of his childhood energy. (The bark on one side of that tree eventually disappeared altogether.) He practiced until he could stick that knife of his into the tree from any position in the yard, standing, walking, or running full tilt.

By sheer luck there were five boys Daryl's age living within two blocks of his house who would become All-State caliber athletes. This "dream team" hunted together, went sledding, and skated on the frozen river. They didn't just skate, they skated to the max—jumped barrels, did anything that would push them to the limit. Daryl began to think sports was his calling—clearly, he was a natural.

In junior high he made first team in football, and was team captain through most of junior high and high school. But his life in sports was dealt a major blow in his senior year when he dislocated his left shoulder. They strapped the offending arm to his body, and he continued to play. But he couldn't play defense because he couldn't reach to the left. He was frustrated, and after the season he went into emergency conditioning; he did push-ups, sit-ups, and lifted weights endlessly. Now surely he was invincible.

Daryl had dreamed of becoming a football coach, but there wasn't much likelihood of his even attending college. His brother Vic, who played the trombone, had enrolled in the music program of a state college ten miles from Waterloo. But there were no funds for any of the kids to go away to school. So Daryl never really thought about it until one day in the fall, several months after his high school graduation, when his girlfriend's sister brought up the subject. She taught math down at Iowa State and was strongly in favor of a university education.

"You know the football coach is my landlord," she told Daryl.

"Archie Steele's your landlord?"

"That's right. Why don't you come down some weekend and I'll introduce you? Maybe he could help you get a scholarship."

So Daryl packed a bag and headed down to Iowa State. The campus was green and beautiful, but the coach was terse, almost curt. "So send me your movies—whatever you've got on yourself." Then he turned away; the interview was over. But Daryl knew coaches—they all had their style. Maybe there was some hope.

A few weeks later Daryl had won an athletic scholarship to start school in the spring. In between, his dad got him a job at the Rath plant as a night shift keypunch operator. There were perhaps 50 people sitting in one room, all entering data about cuts of meat, sales figures, pounds of sausage spice, payrolls. He hated office work, hated being inside, but decided to make a game of it—to see how fast he could get. In about a month it got to the point where people would gather around him to watch; he was a lot faster than any of the other workers.

When the time came for a first evaluation, Daryl felt some anticipatory pride. But when his boss called him into his office, he wasn't smiling. "You know I can't put down on your record what you're actually doing." he said.

"You can't?" Daryl protested. "Why not?"

The boss looked at him sharply; he wasn't used to being challenged. "Well, they wouldn't believe it, for one thing. And for another, they'd start expecting everybody else to do the same."

Whoa, what's wrong here? Daryl thought. But he couldn't say anything; after all, his dad worked there. It was another major learning experience—first an injury had tied one hand behind his back. Now it was part of the rules.

In March Daryl escaped the keypunch slot that would hold some of his co-workers all their lives, and moved down to Iowa State. He wasn't sorry to leave home; by this time the river he loved had become so polluted that practically nothing could live in it. And the way of life he'd learned on his uncle's farm was changing, too. Planes spewed out chemicals onto Iowa's fields, agribusiness was taking over family farms, and the vegetables didn't taste the same any more. Mass production had become the norm.

Daryl focused on his football career. In training sessions the coach proved to be a hard driver. There were hours and hours of grueling practice, and by the time spring football season started, Daryl was sure he was in awesome shape. He could push a two-man blocking sled 50 yards by himself, with the coach on top adding extra weight and shouting directions. Steele, whose look was somewhere between appraising and approving, seemed to have some hope for the Kollman kid.

But it wasn't more than a few minutes into the first quarter of the first game of his first college season that Daryl dislocated his shoulder again, and they had to carry him off the field. "Well, that's it for you for the season." (Archie Steele was never one to mince words.) For two days Daryl mourned, then he made an incredible effort to get in shape so it could never happen again. But when fall came and the coach made him wear a harness to play, he quit football and went out for wrestling. "I figured you don't have collisions in that sport. But then I promptly dislocated my *right* shoulder," Daryl laughs. "I started to get the message and began looking at academics."

This "unplanned experience" (as Daryl likes to call what most of us term "problems") got Daryl out of Phys-Ed and into Pre-Med. It was a natural choice; he was interested in science and was good with his hands. "I could have been a good surgeon, I think," Daryl says. "I really loved knives." For the last couple of years in college he worked in the emergency room at the local hospital, where he saw all kinds of injury and trauma. But nothing bothered him—as long as he could help. After two years in Pre-Med he applied to the University of Iowa Medical School. His grades were terrific, and he hadn't the slightest doubt he'd be admitted.

At that point a second unplanned experience jolted Daryl. He didn't get in. Of course, he had no real way financially to go through anyhow, he realized. Also, at that time there was a definite hierarchy in medicine that didn't easily admit outsiders. So Daryl decided on the next best thing—he switched from Pre-Med to a teaching degree. Then he married his girlfriend Kathe, so by the time he graduated from college he had a wife and a newborn infant, Peter. They talked about where Daryl might find a teaching job.

"California," Daryl said. "Just as long as it's California." Ever since that first trip to the ocean, California had remained a magical place for Daryl. Kathe agreed. It sounded exciting. But where? There were openings for teachers in San Francisco, San Diego, Los Angeles, and Santa Barbara.

Daryl applied for them all, and Los Angeles came through—$5500 for a year's work. They whooped and celebrated. They'd always wanted to live in Hollywood. "So where's the school?" Kathe asked. "What's the name of it?"

"Fremont High."

"Where is that?"

"I'll get the map. Here it is. It's in Watts."

"Is that close to Hollywood?"

They drove out of Iowa and headed for the Pacific Ocean. After absorbing a few realities (like the cost of housing in Hollywood), they rented a tiny apartment in Downey. Daryl would leave in the morning before 5 AM to negotiate the 52 stoplights on his way through increasingly tough neighborhoods to Fremont High, which was perhaps the very worst of the L.A. schools at that time. Fremont had 4,000 students and was 90 percent Black. Daryl had known exactly two Black people in his life, and both were acculturated into the Iowa scene. The kids in L.A. were different.

When the other teachers heard that Daryl planned to do demonstrations and experiments in class, they said, "You can't do lab! They'll tear the place down."

I'm from Iowa, Daryl thought. *I'm tough. I can do anything.*

"They'll be throwing acid on each other. You watch. You'll see."

Undaunted, Daryl prepared to enlighten the inner city kids of L.A. about the wonders of chemistry, biology, and physical science—truths that he would soon learn were almost entirely irrelevant to their lives. Every time he'd begin lecturing, the kids would start talking and laughing among themselves. He'd do demonstrations, set off explosions, and probably a third of them would go to sleep. Almost nobody paid attention—and nobody made much progress.

One day there was a commotion in the corridor. Daryl went out to investigate. There was a pool of blood outside the room, blood along the hallway and down the staircase. Daryl followed as far as he could without abandoning his class; somebody had been stabbed. Another time a fight began outside the class. He waded in and broke it up, incurring the enmity of one of the more violent students. "Next time I see you," the guy threatened, "I'm gonna stick a knife in you." And he sounded like he meant it.

Daryl's habit of getting to school early came in handy about then. One morning he went to the men's bathroom and saw an older Black teacher sitting in the lounge smoking a cigarette. The man had worked at the school for a long time and had seen a lot of teachers come and go. "I've been watching you," he said to Daryl. "I think you're going to need a couple of lessons in how to survive the day. How not to get yourself killed." He showed Daryl how to protect himself when he was in the corridor by walking sideways with his back to the wall. "And another thing," he said. "Give those kids an assignment as soon as you walk in.'

To Daryl, giving kids a worksheet was the last desperate thing you did. But he tried it one day, and it worked. There was instant silence in the class. Almost two thirds of the kids tried to fill out the paper, though their answers often had nothing to do with the questions. And when exam time came almost no one passed. They'd been taught rote, to conform to rules, and that's *all* they'd been taught.

"I can't make a dent," Daryl told Kathe one day. "There's just no way. I want to help these kids so much, but . . . maybe it's just useless. Maybe I need to be teaching where I can make a difference." At the end of the year Daryl accepted defeat and returned to his home town in Iowa.

Watching the Watts riots on television the next year, Daryl saw teachers who'd been dragged through the halls and whipped with chains at Fremont High. He was now teaching at comfortable West Waterloo, where he had a chance to coach several sports, and where most of the kids in his classroom both listened and learned. But even so, he recognized there was a lot lacking in our educational system, even in well-financed school systems like Waterloo.

Daryl won continuing education scholarships at Ripon and the University of North Dakota in physics, chemistry and calculus. From '66 to '69 he taught in a suburb of Chicago, but he still wasn't satisfied. He was looking for an excellence in education he hadn't found yet—a way to transmit not just the facts, but also the essence of learning.

In 1969 Daryl applied for a National Science Foundation scholarship to Harvard for a master's degree in Science Education. Only ten people in the country would be chosen. For several weeks the Kollmans' life revolved around the mailbox. Then one day a fat letter arrived; they'd sent back more than he'd sent them, and so he knew. "We opened that letter with incredible anticipation," Daryl says. "Massive celebration."

He took a leave of absence from his high school teaching job in Illinois, flew out to look at Harvard for the first time, and was shocked to see that there was no campus to speak of, at least not the beautiful green campus Midwesterners are used to. "Anyone who's been to Harvard Square knows that from the cornfields of Iowa there's a lot of difference," Daryl laughs.

The ten successful candidates were shown through the School of Education and then each was assigned an office. They came from all over the country, but soon became a close-knit group. Their faculty advisor, Dr. Fletcher Watson, was well known for his *Harvard Project Physics*. A number of other teaching innovations had recently been introduced, so there were intense discussions and significant breakthroughs on a daily basis.

Strongly influenced by the book *Education and Ecstasy*, by George Leonard, Daryl had been experimenting with new teaching methods for several years. He'd never forgotten those kids in Watts, and his frustration at not being able to reach them; now he had the chance to hone his new theories and develop them. He finished the year vowing never again to be away from that kind of excitement. However, an unplanned experience was about to alter those expectations.

When he got back to Illinois Daryl found that another teacher had been given his old job. Daryl had been transferred to a new school that had just been built. That was okay, he thought at first. He went in a few days early, took his books with him and spent the day setting up the room. But he couldn't see out; there were no windows in the classroom. The administrators had decided that children might be distracted from their studies if they could look out the window.

Daryl went home and thought about it. Then he went back. He spent two more days in that claustrophobic classroom. "Finally," he says, "I realized there was no way I could teach school in a classroom without windows, I just couldn't. I couldn't imprison the kids or myself.

"A friend of ours had a Montessori elementary school. We'd lived with them the summer before. She was always telling me about Montessori methods, but I was letting it roll off my back. Now she said, 'Come with me and I'll show you.'

"I went into that sunny classroom and there was a teacher sitting on the floor. Half the kids were around him and the other half were quietly working at different projects. I stayed for the entire day totally enthralled. I thought, *this is the feeling I want in a classroom.* I'd never found it, never seen it before. But I knew this was the way teachers were supposed to relate to children."

Daryl then committed himself to go to Italy for a year to study the Montessori approach to elementary education. It was a major change for him; till then he had felt that teaching higher grades was more prestigious. The family flew into Luxembourg, then took the train to the medieval city of Bergamo, about 30 miles northeast of Milan.

Daryl was intrigued by the ancient houses with their solemn stone facades, then surprised and enchanted when an occasional open gate afforded him a brief glimpse of the beautiful, profusely colorful gardens inside. But once classes started, there was little time to explore.

"They lectured in Italian and translated paragraph by paragraph. It was intense. They had so much information to give, they couldn't slow down. The American way of learning is to ask questions, but questions weren't allowed here. So in a sense, it was stifling. But it was also a great experience.

"When I came back I began teaching five- and six-year-olds the first year—kids who couldn't get into public school, or who'd been thrown out for one reason or another. Nearly all of them were behavior problems; that was an incredible challenge. It was a class of extremes. One child was a true genius in every sense of the word, and another—all he ever thought about was how to do personal damage to himself. He'd stab himself with a pencil when nobody was looking. I also had my first case of severe nutritional deficiency: a little girl who ate only saltines and butter. They were all in my class. For the first time, I began to see how diet played a part in children's behavior."

Several years passed, and Daryl taught a number of Montessori classes. He was doing what he loved at last—but his personal life was less happy. His marriage had fallen apart.

With a decade of teaching experience by this time, Daryl had become increasingly interested in the correlation between his students' diet and their academic achievement. In the Montessori class everyone ate together; they brought their own lunches so it was possible to see cause and effect very clearly. "It seems to me," Daryl says, "that behavior problems usually stem from learning disabilities. If kids are learning, they don't have as much trouble with behavior. I saw that the kids who were eating sugar weren't able to sit still in the afternoon. They'd have a sugar fix and become so jittery and edgy they couldn't learn—then they'd come crashing down."

Daryl understood that syndrome; his mother had always equated sugar with love. He remembered her pies and strudels at the holidays—and how bad he used to feel after eating them. So sugar was a problem, and so was junk food. But Daryl was beginning to feel that the decreasing quality of our soil, and therefore most of our food, also played a role.

He sometimes talked to the parents of his students about what he observed. One of the parents who helped regularly in the classroom also saw the connection between diet and behavior. She was the mother of a very bright little boy named Sevin Straus, and she impressed Daryl with her patience and ability with the kids. "Children were all of my life when I met Marta, and I saw that she shared that value in very special ways."

Marta understood the Montessori precept that the environment should be as beautiful as possible to encourage learning. When the budget didn't provide for them, she made flashcards, drew posters and banners, all with the flair of an artist—colorful, cheerful, and inviting. Daryl found himself increasingly interested in this high-energy and inventive young woman.

He was beginning to fall in love.

CHAPTER 6

Celebration Partners

"I'VE ALWAYS THOUGHT OF MYSELF as incredibly lucky," Marta Kollman says. "If everything fell out from under me tomorrow I couldn't complain. I've had an enchanted life, with the opportunity to do so much. I was brought up by parents who believed I could accomplish anything I wanted to do, and who loved each other very much. My mother was unique for her era, the original health, ecology and conservation sort of person. When my father would came home at night she'd be in the kitchen. He would hug her—he always called her 'Skinny'—then he'd reconnect with each of us. It seemed to me that I always delighted him. I grew up in a neighborhood where none of the moms worked and divorce didn't happen. I could walk into anybody's house and anybody could walk into ours. I was totally at home in all these different places; it was unique, I don't think that happens any more."

Marta and I had kicked off our shoes and were sitting on stools in her kitchen, which, like her bright, cheerful office, is covered with mementos and pictures of all the people, especially children, who mean a lot to her. Their home is situated on a hill where Daryl can climb up a mountain straight out of his door. From the front of the house there's a stunning view of Mount Shasta. And there's a swimming pool in the back. When Daryl and Marta first moved to town, they rented a house in a nice but distinctly middle-class neighborhood. Several years later, when a home nearby became available, they bought it rather than going upscale—despite their soaring success in the interim—because it would enable their son Justin to stay in the same neighborhood with his friends.

Limited by a small lot, they've made enormous changes in the structure of the house, the most striking of which is the huge new kitchen, with what looks to be a 20-foot ceiling, and the most scrumptious commercial stove I've ever laid eyes on. This room is the one to which all guests not only gravitate, but stay; there's no thought of sitting in another room. Kollman get-togethers are informal, ranging from do-it-yourself pizza to fondue made by their son Joe, with everyone pitching in to help. The Kollmans don't put on airs.

Discussing the pivotal moments in her life, Marta looks at each from a positive point of view. She'll fix you with her stunning smile and tell you how each experience has turned out to her advantage—it's all part of that unfailing optimism. Her earliest memory, for instance, was that adults were always glad to see her. "It was a sort of an 'I must be okay' realization," she says. "People didn't just say Hi—they stopped and talked. They invited me to go places, and what that said to me was that my parents must be nice people, too."

Marta's dad was an executive with Sears Roebuck; her mother became an archæologist later in her life. Marta was raised in the country in Inverness, Illinois, northwest of Chicago, in a white Cape Cod house with green shutters. Every house in the area had at least an acre of land, and often a lot more. Marta grew up a tomboy. "When other girls were learning to curl their hair and do that tricky stuff, I was riding my horse, or climbing a tree, or seeing if I could hold my breath longer than anyone else."

Marta had an older brother Jim, who is now Cell Tech's full time ecologist, and an older sister (now an Episcopal priest). There were no kids her own age to play with, so she had to plug herself into a group of kids as much as eight years older. "I don't know when I realized it, but all I had to do was just be around them and absorb it all. Things were always going on around me lickety split. Do you know what an *advantage* it is to be the youngest?"

When Marta was ready for first grade, her parents sent her away, an hour by public bus, to a Catholic school in Barrington, where the townies lived completely different lives from those of the friends she was used to. "They walked on sidewalks, they listened to records, they weren't outdoorsy at all."

There were adjustments to be made at school as well. Marta jokes that, "For every aberration in my personality, I can point to the nun who's responsible. I'm not a blind faith person and many of the things they were trying to teach me didn't make sense. The nuns used to bribe me with holy cards not to ask difficult questions, and I worked it up to a statue. Because I used to ask questions like, 'Sister, God is all-good and all-forgiving, right?' 'Oh, absolutely,' they would say. And I'd say, 'Then there can't be anybody in Hell.'

"I never got along with the sisters; they were always trying to suppress me, which only made me more doubtful about what they were up to." She flashes that brilliant smile. "But then I went to public high school, and that was great!" (Marta is one of those rare souls for whom her 25th high school reunion was an utterly positive experience. I envy her that.)

In her senior year, Marta suffered two major unplanned experiences that were about as devastatingly serious as any 17-year-old could imagine. One evening an old boyfriend picked her up in his family's Chevrolet, and they were headed down the road to the Dairy Queen when a car traveling at high speed veered into their lane. The resulting head-on collision nearly killed Marta; she was unconscious for five days. "They didn't even bother setting bones the first night; they just tried to stabilize me. People who stopped at the accident and who'd known me all my life didn't realize it was me."

Nine doctors hovered over Marta that night. They came out and told her mother that she wasn't going to make it. "Then they said that if I ever regained consciousness, I'd be a mumbling idiot." She laughs. "Mom and I used to kid around later, saying, 'Well, they were right.'"

When Marta finally came to, she had almost no memory. Her best friend Kathie Sheldon made a huge get-well card and everyone at school signed it. (Marta had it framed; it hangs on that 20-foot-high kitchen wall). "Kathie came to see me every single solitary day, and she was the only one who wouldn't humor me. I'd say to her, 'So when did Bill leave for school?' (Bill was her boyfriend and he was leaving for Dartmouth.) She'd say, 'He left Tuesday.' And I'd say, 'Oh.' Then a few minutes later I'd ask her again. She'd say, 'I just told you—now before you ask me a question again, think! Ask yourself if you just asked me that same question.' And then I'd do it a third time, and she'd say, 'No, no, no. That's three times in a row.' When I finally got back to school, I'd read a book. I'd understand it, close the book, but a moment later I couldn't remember if I'd ever opened it. My mind was like a sieve."

After the accident, Marta had to undergo five plastic surgery operations. I asked her if she ever worried that her beauty might be gone forever. "No," she says. "I never thought of myself as a beauty. I just thought I was lucky because one of the best plastic surgeons in the country happened to be on call that night. It could have been my whole face, but it was just my forehead, and I knew I could always wear bangs." She looks at me and laughs. "I don't mean this in a flippant way, but at least it gave me something to talk about. Nobody would believe it who knows me, but I'm basically shy. I have a hard time with small talk and if there's a lull in the conversation it makes me nervous, so I start talking like mad about almost anything just to fill in. After the accident, if there was nothing else to talk about, I could always launch into what had happened and what I'd learned from it."

Marta still doesn't remember the accident itself. I know that's not unusual, but what is unusual is that she harbors no bitterness about it. Unless she's asked, she doesn't even mention that it was someone else's fault. It's clear that in Marta's life, nobody has done her wrong; she takes responsibility for all that happens to her. In fact, like Daryl, she tends to take advantage of it. Perhaps, as they both believe and demonstrate, it's not how well we plan our lives that makes or breaks us, but what we do with the unexpected things life hands us that counts.

The following spring the second blow struck for Marta. She was beginning to catch up in her studies, and was across the street with her neighbors' little boys. Their parents had gone to Fort Lauderdale and had hired Marta to babysit. "They'd given me their Buick Wildcat to drive while they were gone, and I was going to stay at their house. I was just settling in when Mom called and said that Dad had died of a heart attack. It was a total shock, completely unexpected. Mom said she'd call the neighbors and have them come right home. They came back that day."

Marta is quiet for a moment. Then she says, "He died in March." She looks up at a watercolor on the wall. "I used to love to watch him paint. He had such beautiful hands. That picture of the church over there is my Dad's. He was so artistic; I think that's where I got my feeling for color." She wonders aloud how her life would have been different had he lived. But it happened, that was it. Again, as she finishes the story, her focus is on the positive. "It was a horrible, lousy, cold, rainy day for his funeral, and I thought that was just perfect—I didn't want it to be a nice day."

Since she had always been interested in a business career, after graduation Marta went off to the prestigious Katherine Gibbs Business School in Boston. Back home again one day, she took her ailing Austin Healy to a shop—till then *nobody* had been able to fix it—and that's where she met her future husband. Jim Straus was a mechanical genius—he could make anything run. Because of Jim, Marta got interested in motorcycles. She started riding, and eventually began off-road racing as well. "I was pretty good at it; I even won a trophy," she says. But didn't she have any fear? (I would, if I'd ever had an accident like hers. Actually, I would anyway.) "Oh gosh, no," she says. It's clear Marta Kollman isn't afraid of much of anything.

"After we were married," Marta continues, "a friend told us about a motorcycle dealership for sale in Skokie, Illinois. We looked into it—it seemed to be a good deal, and I had some money my dad had left me. We decided to take a chance and buy it." So by the time she was 20, Marta owned her own business and was beginning to gain experience.

When she was pregnant with Sevin, their first son, her husband worked six days a week, sometimes 12-hour days, and their business thrived. The marriage was successful, too, for a time. Marta had a second child, Justin. There were three pregnancies between the two boys, but again Marta skips quickly over the unpleasant part. "Jim was the first person I'd ever considered marrying and I still think he's one of the best people I know. But somehow our marriage just didn't work out."

When Sevin was old enough, Marta placed him in a Montessori school. A couple of years later, she met Daryl when he became Sevin's teacher. They fell in love, and were married outdoors in the park at The Netherlands Carillon in Arlington, Virginia. Two close friends were the only witnesses. "It was gorgeous," Marta says. "You know how manicured everything is there." I did—I used to live in D.C. "And the carillon rang before the ceremony, and it rang after—which was perfect." They had a very brief honeymoon at the Hay-Adams across the street from the White House.

Daryl already had two boys, and so did Marta. Mixing their two families was a challenge, but for the most part it worked well. "I was happy to have boys," says Marta. "Girls seem too complicated. For instance, with the boys, when I'd launch into my 'I'm not your slave' speech, they'd just stand there and nod, trying to look serious. You could see them thinking, 'This is lecture number 43 and it lasts about 2.3 minutes. She's Mom and we love her, she's a female, she's a little bit nuts, but this will pass.' Then they'd go back to what they were doing. Girls, I think, would sigh, 'How could you *say* a thing like that?' Then they'd go off to their room and not talk to you for two days."

Daryl's boys got along with Marta, and Marta's with Daryl as well. When I interviewed him, Justin Straus told me how great it was to have Daryl for a dad. "My very first memory of him was a physics lesson he gave me. I must have been about three. He was riding me on his bike down a hill, and Joe and Sevin were racing us on their bikes—he said we'd win because we had more weight and so we could go faster. He was right." Justin smiles as he remembers.

"Being married to Daryl is incredible," Marta says, borrowing her husband's favorite word for a moment. "I never get tired of being with him, I love hearing the garage door open, knowing he's come home. Daryl has made me laugh harder than anybody I've ever met. Most people see him as a friendly, caring, sincere man who's saving the world, and he is—but there's also a side of Daryl that's just the funniest person I've ever met. Daryl does imitations—he'll imitate inanimate objects and make them come alive, or he'll string together words I would never have thought to put together. It takes my breath away and I start laughing so hard I can't stand up.

"He does all these goofy things that melt my heart. We both love being in water—swimming and snorkeling. I don't know what it is, but the combination of Daryl and water . . . I honestly think I'll drown laughing—I just can't help it." Once launched on the subject of Daryl, Marta can't stop. "He has a very organized mind. He's taught me a lot, but mostly I just have a crush on him—I think he's amazing, so multi-faceted. He's romantic, loving, he loves to dance—and that cracks me up; he's good at it but he doesn't even care if he's good. He just likes it."

After they were married, Daryl accepted a job in Hilton Head, South Carolina, where he was Headmaster of the Lower School at Sea Pines Academy. During one summer vacation in the early seventies, the family drove out to New Mexico to visit Daryl's older brother, Vic. Vic had left his trombone and musical career behind to go into science. Now he was part of a government team at Los Alamos that was cultivating algae, tobacco, and other fast-growing plants in an atmosphere of Carbon 13, for biomedical use.

"Carbon," Daryl explained to me one day, "usually has twelve electrons, but some carbon has an extra neutron in its nucleus. If a patient ingests a food that carries this isotope, it can be traced through the body to get information on how various organs are functioning. And because it's not radioactive it doesn't cause any harm to the body."

As his brother showed him around the lab and shared with him the work that he was doing, Daryl immediately focused on the experiments with algae. Since little was known about the long-term effects of that particular algae on humans, it was being fed to laboratory rats to test its safety. Confined in small cages, the rats were nervous and high strung; they often cannibalized their young. But in the animals fed algae, the cannibalism had inexplicably stopped. And the rats had become healthier.

When Daryl saw the results of the rat studies, the only conclusion he could draw was that the algae must contain some beneficial nutrient that had been missing from the rats' diet. And he couldn't help but correlate what was happening with the rats with what was happening in his own classrooms. By now he was convinced that the number of kids unable to focus or concentrate was increasing. He went back to Hilton Head for another year in education, but couldn't get the algae out of his mind.

The next summer when the family visited, Daryl and his brother went to the computer at Los Alamos, which at that time was a gigantic piece of equipment hooked into all of the data banks available to the federal government. When they asked the question, the computer confirmed what Daryl had begun to suspect—that algae *was* the most nutritious food on the planet. Excited, they asked for a complete print-out of all the available information on algae, and generated a stack of papers about a foot and a half tall.

During the next year Daryl and Marta read through all the material and were astounded by the scope of what they were learning. Marta saw a phenomenal market for algae—as human food, animal feed, and fertilizer for depleted soils. People in Japan, they learned, were already eating algae, not just seaweed, but the microalgae known as chlorella, and they seemed to be receiving benefits from it. Marta said, "Let's grow algae commercially. Let's start a business."

Vic and Daryl thought it was a good idea, and the three of them talked about it a lot. But as Daryl says, "Talk is about all it ever would have been, if it hadn't been for Marta. It's always been Marta willing to try the big things—the big adventure. So in that sense, she's the visionary. After all, Vic and I grew up in the same family; our dad never tried anything new, and even though my mother wanted something different, she didn't have the means to make it happen. I doubt if I ever would have gotten into algae without Marta's courage, because I had a picture of my life involved with teaching."

Victor applied for and received a grant to explore the commercial feasibility of large scale algae production. With the promise that they could be involved, Daryl and Marta decided to make the leap—he would quit his job and they'd move to New Mexico to work with Vic on the project. Unfortunately, before they could actually get to New Mexico, most of the grant money had already been spent. Vic had constructed an algae pond out in the desert, but so far no algae had been successfully grown. Still, even without the guarantee of any income, they packed up and headed west.

Once out in New Mexico, Daryl and Marta set up camp with their boys in an apartment in Albuquerque. A 4x6 piece of foam rubber leaning up against a wall served as a sofa. Marta's mom gave them a dining room table, but except for beds, that was it for furnishings.

Vic had built a prototype algae pond at the University of New Mexico Agricultural Station in Los Lunas. The Kollman brothers worked hundred-hour weeks for months to get the project up and running so they could produce samples and complete the grant. After school in the afternoons Marta and the boys would join them and they'd all work together. It was a painstaking process; the algae was extremely temperamental. It had to be fed exactly the right nutrients and kept at a precise temperature. "Sometimes it would just up and die," Marta remembers. "It would be green one moment and the next morning you'd wake up and it would all be dead." At last they succeeded in growing enough algae to complete the grant.

"One day there was a letter from Japan," recalls Daryl, "and we answered it. They sent representatives over to talk to us about developing large-scale facilities for growing algae, and we thought we'd hit the jackpot. We began looking for land, and found a piece of uncleared desert. It was 54 windy, sandy acres with cottonwoods along the river. Though it was basically arid ground, you could dig down just three or four feet and hit water. And it had never been used for agriculture, which was very important."

They bought the parcel, and Marta's mother provided the down payment for a double-wide trailer which they moved to the land. They began working around the clock, cutting trees and burning brush to prepare the land for the construction of plastic-lined ponds. It was rough work in 100-degree heat. They'd end the day with smoke-blackened skin and sore muscles, totally exhausted. The plan was to build 50 acres of ponds.

"We signed a contract with the Japanese company to go from zero to 30 tons of algae per month, within a year. It was going to be 50 tons a month the second year, and 100 the year after that—an enormous contract. We made millions of dollars—on paper."

Construction began. "We moved out of the apartment in Albuquerque and down to Bosque," Marta recalls. "Then we built the big pond that was covered with fiberglass, and it looked like a frosted white spaceship at night because it was lighted to help the algae grow. That's when Harrison H. Schmitt, a U.S. Senator from New Mexico at that time, saw a picture and came down to see it. Schmitt had been an astronaut; he was the first non-military person to walk on the moon."

As a result of Schmitt's visit, reporters called and asked questions, and a couple of national stories were run about the Kollmans' work on the potential of algae as a low-cost source of protein. One of the articles included a picture of a scale with beef on one side and a small amount of algae on the other, a picture which was eventually seen by people all over the world. In the small town of Klamath Falls, Oregon, an attorney read the article and was reminded of the algae that grew wild in Klamath Lake. Wondering about its potential, he wrote to the Kollmans, telling them there were about a zillion tons of it, but he didn't know if it was the kind that would be useful or not. Daryl immediately called him up and asked him to send a sample. None arrived, however.

With their multi-million dollar contract to cultivate algae in hand, Daryl and Marta went to their banker to obtain financing for the construction of algae ponds, but were turned down. "We were so naive about the money system," Daryl says. "We believed in the myth of America as the land of opportunity. But the system isn't designed to help people, it's designed to keep them where they are. Through '76 and the first part of '77, we talked to just about everyone we knew." Both Marta and Daryl thought they had some pretty good investment contacts, Marta through friends of her late father, and Daryl through people he had known at Hilton Head. But it didn't seem to help.

"We learned a lot, we told our story over and over, but found that loans aren't usually given to people who need them. We also learned that Small Business Administration loans aren't really intended for small businesses. It was a real eye-opener for us to find out that nobody lends you money unless you already *have* money.

"We weren't willing to give up, though. It came down to the realization that we weren't going to be able to move forward unless we risked all of our own resources. So Marta and I made that commitment together. It was total and absolute; we knew exactly what we were doing. We sold property, we sold a car, we sold everything we could to get the money for the materials to build the ponds. We went into massive debt—all based on our belief in the future of algae in the human diet."

CHAPTER 7

Difficult Choices

DARYL AND MARTA WERE LEARNING how almost impossibly hard it was to grow algae. "At the time, we had a 3,000-gallon pond and a 10,000-gallon pond, and were gradually moving toward building a 100,000-gallon pond. When you're growing algae," Daryl explains, "you start with ten milliliters of water—a very small amount, maybe two tablespoons. Then you multiply by ten, and by ten again, and keep upgrading. And at any step along the way you can fail. It can simply die. You can't just put a few cells into a massive body of water—it doesn't work. We also learned that there's a lot of guesswork involved in coming up with the perfect proportion of nutrients to make the algae flourish." In essence, the Kollmans were doing what farmers do when they add fertilizers to grow field crops (though the latter tend to be hardier). It was an artificial situation at best. But there didn't seem to be any other way.

They continued to work on building algae ponds with whatever funds they could put together, but to earn a living Daryl also worked with Vic on a number of other projects. There were so many potential uses for algae. They traveled all over the country talking about it and promoting it. Finally they landed a grant for a demonstration project using algae to clean up brackish water in Lumberton, a little town with great Mexican food and one bare-light-bulb-and-a-bed hotel, in northern New Mexico near the Apache reservation.

The town's water supply came from an underground aquifer where there was a lot of coal. Tests showed large amounts of methane and other gasses. So not only was the water undrinkable because of its salt content, but you could literally light the faucet with a match, there was so much dissolved gas in it. Residents of Lumberton would routinely shock visitors by doing exactly that.

At that point, Daryl and Marta pulled in Marta's brother Jim Carpenter, a contractor in Colorado. Daryl, Vic, and Jim designed and built the system themselves over the course of a summer, fall, and winter—again it was exhausting physical labor. Daryl had essentially moved up to Lumberton by that time; Marta and the kids visited when they could. When funds ran out, they added some of their own money to complete the project. "It was that important," Daryl says, "to make progress. I'm a progress fiend."

They laid pipe to run the water from the town well into a solar greenhouse. There a special kind of algae would be grown in ponds to absorb the gasses, salts and other pollutants. Then the purified water would be run through a filter to screen out the algae, before being pumped into the town's storage tanks.

The Governor of New Mexico came on opening day and they filled the greenhouse ponds with the noxious water. For the grand ceremonial opening, instead of cutting a ribbon, they dumped a 55-gallon drum of algae into the tank and it just disappeared. A few hours later there was a tinge of color to the water, and soon after that it was solid green; this particular algae divided about every 20 minutes. Almost overnight it went from a pond full of polluted water to a green biomass.

When they turned the system over to the city of Lumberton, it looked as though algae might facilitate a new future for agriculture in New Mexico. There were huge quantities of brackish water just below the surface over much of the state that could have been purified and used to irrigate crops. But although it had a lot of potential, for a number of political and logistical reasons this successful pilot project was not repeated elsewhere.

After Lumberton there was another grant to remove toxic residue from feedlots, using algae. Daryl explained to me that nitrates from feedlots are often run directly into rivers and streams, damaging the entire ecosystem. In the course of that project he found himself wading in chicken manure, and saw for himself how factory-farms are run. "Terrible things happen to the animals we raise for food," Daryl says in a rare allusion to the negative. After the feedlot project they got another grant to treat sewage.

Between projects Daryl went out on the road, lecturing and trying to obtain additional grants. So his work on the different applications of algae was providing a living for the family. However, his real interest remained with algae as a food product. While he was traveling, Marta stayed out on the land, and watched over the new algae batches in the lab. That always made her nervous. As she puts it, "The algae seemed to know when Daryl was gone." She also worked in the makeshift office typing the grants, and tried to keep a handle on their shaky finances. Sometimes she took the boys, each on his own motorcycle, out riding for miles across the mesa. It was about the only entertainment available.

Naturally, Daryl was out of town when one day, out of the blue, the "sample" of wild algae from Klamath Lake finally arrived. Marta relates that a family just appeared one day with five 55-gallon drums packed in dry ice in the back of a pickup. She was floored. No way did she have a freezer that big. How to preserve it? She drove into the nearest town and asked the manager of the local grocery, "Could I put these drums of algae in your freezer, please?" He said, "Sure." So they stacked them in there until Daryl and Vic got back.

"When Daryl came home and we looked at the new blue-green algae together, we were impressed and unimpressed at the same time; because we didn't know what it would do,"

Marta recalls. "It was like a total stranger being presented to us; it couldn't talk, couldn't tell us what it was, it had no way to get our attention. So we just kept it in the freezer. We didn't forget about it, but we were unimpressed."

A couple of months later Marta reminded Daryl about the new algae, and he agreed it was time to find out what, if anything, it had to offer. First he tried every feasible means to process it: drum drying, spray drying, fluid bed drying, solar drying. It didn't work. The stuff was beautiful but it was different from what he was used to—it formed a rubbery sheet over itself and wouldn't dry underneath. Stumped, Daryl tried a big grain dryer. The algae didn't rot, but it didn't dry either. Finally, rather than throw it out, he decided to run it through somebody's freeze-dryer. Then he brought the finished product home.

Daryl was in the habit of being his own guinea pig, so he dissolved some in a glass of water and drank it. He'd had some experience with microalgae (spirulina and chlorella)—with chlorella he'd been able to tell some difference in the way he felt overall, but he'd never been overly impressed. "I had no concept of what this algae was at that time. It wasn't in anyone's reality. There wasn't even anything I could read about it." At first Daryl and Marta weren't even certain of the species; they knew it was a blue-green algae, but thought it was *Anabena.* It turned out to be *Aphanizomenon flos aquae.*

Almost immediately after Daryl drank that new algae in the glass of water, he began to feel its effects. "All my life I've kept my eye on my 'energy meter,' because every day that I live I want to live to the fullest. I want to give it everything I have. I've always paced myself so I'm totally exhausted at the end of each day. I'm always aware of it, always watching this 'meter.'

"The first day I ate the new algae, by lunchtime my meter hadn't even moved—I had just as much energy as when I woke up. The afternoon progressed and it went down just the smallest amount. That was important to me; I was working hard during those years because there wasn't any money. Sometimes we worked around the clock, and we always worked around the week, because if we stopped, that was a day without potential. We had a very small amount of resources left and we were watching them dwindle away at a frightening rate.

"At the end of that first day, I still had a lot of energy left, and I told Marta, 'You've got to eat some of this stuff. This is different.' So Marta ate some at dinnertime.

"By ten that night the algae was out of my system and I went to bed. But Marta had taken it at six and couldn't sleep. She lay in bed for a while, then got up and did everything possible in the house—rearranged shelves, cleaned closets, read, wrote—she was up all night!"

Marta recalls, "It was like going from kerosene to jet fuel."

Daryl and Marta started eating the algae on a regular basis and began experimenting with ways to apply it. The benefits seemed endless. Daryl remembers, "We cooked with it, put it on cuts, on the kids' chicken pox. We did everything possible with it. I used to have incredible allergies and would get cortisone shots three or four times a year, but gradually the allergies disappeared, and I wondered where they went. Also, after a while I realized I could cut and scratch myself and not get infected anymore.

"I felt like an explorer of some kind, pushing into new areas. It wasn't even like landing on foreign soil, because we had no expectations at all. We knew nothing about this stuff, no one to our knowledge had ever eaten it. All I could think about were those old stone houses where I used to live in Italy that were connected to each other. From the front they looked like nothing at all, but then I got to know a few people and when they'd open the front gate, their gardens would just knock your eyes out. Behind the worst-looking door would be the most colorful garden you ever saw."

The new algae had a lot of potential, but unfortunately it wasn't bringing the Kollmans any income. There was one brief flurry of excitement when they decided to take some to the National Health Federation show in Los Angeles. Daryl, Marta, and the kids hand-packed enough capsules to fill 200 glass bottles. "Our kids learned to count to a hundred better than most," Marta laughs.

"Marta typed the labels," Daryl recalls. "And then we put tape around the lids to be really professional. The show was at the L.A. Coliseum, which is enormous. We were assigned a booth about the size of a card table."

Across the way was the gigantic and elegant Perrier exhibit. "We set the bottles of algae on the table and stood behind it. We had no literature, nothing. People started to walk by, typical eccentric Southern Californians. They'd look at us and sometimes they'd stop and say, 'Like wow—I don't know what you've got there, but there's sure a lot of *energy* coming off that booth!' But they didn't buy anything.

"Eventually we started giving out capsules for people to try, because we didn't have any other way to interest them in the algae. And we hadn't known enough to bring water for people to take them with, so we sent them over to Perrier. They'd take the capsules we gave them, be gone about 30 minutes, then come back by the booth and say, 'What was in that stuff?! I feel—something happening—I feel great!' We sold out the first day after we started giving samples."

Daryl and Marta soon learned that once people tried the algae, they usually bought some. But it still seemed impossible to find investors. "In those years when we were using up all our contacts to try to borrow money, over and over again we were told, 'You know you really ought to change the name—call it liquid protein. Nobody wants algae. And the color—can't you bleach it or something—get rid of the bright green?' But we stuck to our principles of being up front with it. It was algae and we'd call it algae; we wouldn't try to disguise it."

By now Daryl had become convinced that the wild algae was an infinitely better nutritional product than the artificially raised kind. So during that spring of '78, he and Vic designed and built an algae harvester. Then they sent a technician to Klamath Falls to gather some of the algae from the lake; it took the very last of their money to do it. They ended up with eight thousand pounds of wet algae in the freezer and nothing else. There were no resources left to process or promote it.

In addition, the Kollmans had reached a personal impasse; Marta felt that she could no longer work with Victor. "It's my opinion that he just didn't have the same principles that we did, and I was convinced that the situation would never change." She was also concerned that her children weren't getting the education they needed. The youngest, Justin, was in a woefully bad first grade class. Marta went in every day to help, but at six Justin was so far ahead of the other first graders that he was often teaching the rest of the class, who were mostly poor farm kids.

Finally Marta made the difficult decision to return to her home in Chicago. She hadn't stopped loving Daryl, and was no less committed to the new algae, but as she put it, "I guess it's my nature to keep an eye on the heart and the kids. I wanted more for my children than I was providing, and I was stopped from it. I was responsible for three people; Daryl was an adult and I figured he was responsible for himself. But as a mother—the person who needed to establish the children as responsible productive individuals, I couldn't do it under those circumstances. There just wasn't anything more I could do."

So Marta took the two younger boys and left.

Chapter 8

Lonely Times

Marta was gone, the two younger boys were gone, all the money was gone, and shortly after that Daryl and his brother had a falling out and Vic went off to develop the Klamath Lake algae business by himself. So Daryl had not only lost the woman he loved, he'd lost the algae as well.

So there he was, out in the desert, broke, in a trailer on 54 acres of blowing sand with a 13-year-old. (Joe had elected to stay with his dad.) Their chief assets were some half-built algae ponds that would never be used, and a large trash bag full of Klamath Lake algae capsules.

"Marta left just as the spring of '78 was turning to summer, and I understood why she had to go. She had her limits. Then Victor left in the fall of that same year when he negotiated an exclusive lease on the lake where the algae was. But I couldn't bail out because I was learning so much, studying the algae and its properties. By then we knew the incredible benefits. You'd look at a single facet of it and say, 'Wow! That's neat.' You'd work with that facet a bit, then see another one and say, 'Oh, my God, *that's* possible?' Then there's another facet, and another, and all at once you realize there must be a whole iceberg below it. I couldn't stop. I knew this was what children needed—all the children I'd been unable to help for so many years. The algae had a massive hook on me.

"My brother and I had stopped talking to each other. Marta and I hadn't stopped talking, we talked every day on the phone and we got together whenever we could, but there was a lot of frustration, a lot of hurt feelings, and it ended our marriage. There was incredible pain—awesome pain. But even though there was personal and financial trauma, the centrifugal force of that learning curve helped to keep me in place while everything else was going on.

"Joe and I tried to make opportunities out of anything and everything. He was going to school and working with me too—we did a lot together. We had an unforgettable two years; scary years, though, because there wasn't any money.

"There was an older guy who went through our lives at that point. He was successful, and he'd developed a machine that produced ozone. So he and Joe and I started manufacturing ozone generators, and I looked for potential applications. I talked to the city of Albuquerque and we installed some ozonators in their sewage treatment plant. (If you've ever wondered where the water goes when you flush your toilet, I've been there. I've waded in it; I know where it goes and how it's disposed of.)

"I also did some work with ozone for a dairy farmer in Colorado who'd started a business using chicken manure for cattle feed. I realized later that an understanding of nutrition and animals and ozone—each of these things—was essential for what was to come later." It's clear that Daryl equates unplanned experience with learning curve.

"We installed an ozone machine for that dairy farmer and treated the chicken manure with ozone—took all of the odor out of it—it smelled like fresh hay at the end." Daryl laughs. "It was during that time that I almost went blind looking at the ultraviolet light in the generators. I knew it could be damaging, but I hadn't had much experience, so off and on I'd look at the light. Then one night I went home and I couldn't see. I'd sunburned my eyeballs. The next day I bought protective glasses for me and for Joe. Ozone became such a part of our lives, I think I can still smell it at about one part per billion.

"Joe learned a lot in those years too, from age 13 to 15. He was on the learning curve with me, and we spent a lot of time together because we were way out in the country and there wasn't much else for him to do." Joe points out that since his school in New Mexico was 90 percent Spanish speaking, he stuck out like a sore thumb. "I was never in a fight in school but there were probably ten to fifteen fights a week, and this was only ninth grade. It was macho-ruled and I was a musician; I played in the band and kept quiet. In the afternoons I rode my motorbike by myself; later my dad got us boxing gloves, and we used to spar every night. Until I accidentally knocked him out.

"I can remember being 13 and putting the ozonators together, and having some idea about a way to do it that I thought was better. Dad would listen all the way through, and if he liked it we'd try it—he never treated me like a kid, or as if kids weren't smart. Anyway, we built a machine for that chicken farmer in Colorado that was supposed to separate the solids from the liquids. Dad and I put it together and drove it up there.

"But when we put the liquid in it, it didn't work. As soon as it started up we both realized this huge design flaw in the belt system, so we packed up the truck and headed home. Still my dad didn't get upset. What I learned from him was that you don't have to get mad about something when it's just information; it might take ten or twenty steps to make something perfect." No doubt that lesson has been valuable for Joe, who is now in charge of harvest and production. Over the years he's helped Daryl design and build most of Cell Tech's complex processing equipment.

After two years, Daryl and Joe had worked their contracts to the point where nothing more was happening—there was no money coming in. "We had to sell something," Daryl told me, "so we sold the land. It had increased in value because we'd improved it. So Marta got back the money she'd invested. Joe and I took our share, paid our bills, and got out of debt."

Then Daryl pulled out the atlas; he and Joe looked at all fifty states and talked about where they might want to live. Naturally Daryl wanted to settle somewhere near the ocean. They thought about Florida, the East Coast, the Gulf Coast, but finally decided on San Diego. By now Joe was 15 and ready for high school. They packed up and headed west.

"I substitute taught in the spring of that year for the San Diego School District. I really didn't want to get back into education at that point, but I needed something. By the summer of 1980, I was looking through the newspaper for just about *any* job. We had about a month's worth of rent money left.

"Desperate, I answered an ad for a solar engineer—I knew I wasn't qualified, but I went to the interview anyway. It was a company called Sun Tech. The sales manager, whose name was Ray, told me they were selling hot water systems for homes. He showed me around, introduced me to the owner, and we got into a conversation that lasted several hours. They got excited about me, and I got excited about them. Ray said, 'There's no place for you in engineering right now, but why don't you train to be a solar salesman?' I had no sales experience, but I didn't have much choice. It was a job.

"They had a phone room that generated leads for us, so there was never a cold lead, but some of them were pretty darn cool—and you didn't make any money until you made the sale. There was intense pressure from the managers to make sales." Daryl explains that he didn't particularly like the way the company handled its sales; he thought it wasn't necessary to pressure people the way they did. So he didn't use pressure—and he tried to influence the other salespeople as well. "I tried to bring integrity into their day. The experience was valuable; it led to the way we do things in sales at Cell Tech. I learned you could do sales with integrity or without."

While Daryl was working at the solar company, he began to read about the growing number of homeless and hungry children in the world, and decided he wanted to help. "I was also trying to get out of sales and into a job that would be more meaningful." Using the experience he had in writing grants, he prepared a proposal for an idea that he called *Professional Parents*. "I wanted businesses to fund a combination school/living/work situation for inner-city kids. I sent the proposal out to all the major industries in San Diego and I think I got one reply. The whole effort was totally nonproductive.

"I began wondering why people with good ideas aren't always successful at making things happen—I was feeling pretty down about it. So one day I went out and sat on the pier in San Diego and watched the waves roll in. There was so much harmony out there. I sat and analyzed my motives. I knew where my heart was—with Marta, and with what I wanted to do for children. At that moment I felt nothing about making myself wealthy or important or even without pain; I just wanted an opportunity to help kids and I kept asking myself, 'Where is it? Where is the money to help kids?' But I didn't get any answers.

"That night when I went to sleep I dreamed that I saw the Goodyear blimp flying overhead. It had a clear message on its side. The message read: 'IT'S—IN—THE—WATER—.' I sat up in bed and instantly made a clear association with the question I'd asked earlier in the day: *Where is the money for the kids?* The answer: *It's in the water*. The only thing I knew about that was in the water was the algae, so I clicked on that immediately: "Oh, my God, it's the algae!" That night, alone in the dark, I totally committed myself to doing everything within my power to move the algae forward; it was a total, absolute commitment—everything else fell out of my life.

"I kept working at the solar place because I had to pay the bills. But I also went into high gear promoting the algae. The only people I knew at that point who had money to invest were the owners of the solar company. I didn't share my algae with them because I didn't have enough to share—I just told them about it. They were interested; their own business was going downhill. Recent government regulations had given regulatory power to utility companies, who of course had no interest in promoting their chief rival, solar energy. So together we began to think about a strategy to harvest algae from Klamath Lake.

"In the spring of 1980 the four of us formed a partnership—getting ready to do what we called an assault on Klamath. I didn't know how we were going to do it, but I knew we had to do something—because I was running out of algae. In the bottom part of our refrigerator there was a plastic bag that held the last of our algae capsules. I doled it out each morning—four for me and four for Joe—but the bag was getting smaller every day.

"One Saturday I'd taken time off to go roller skating on the boardwalk—when suddenly I saw my brother Victor and his girlfriend standing there at the side of the path. I hadn't talked to Vic for a year and a half; he was the last person I expected to see. I spun around and stopped. He said, 'Hi,' and we talked for a few minutes. Then Vic asked, 'How would you like to harvest some algae from Klamath Lake?'" Daryl laughs uproariously as he remembers. "*Did I want to harvest algae?!*" He laughs again. "What a concept!"

"Vic wasn't selling much algae at that point. And here I had Ray, a super salesman, and three other investors ready to go for it. I outlined to Vic what we wanted to do—set up a company to purchase and market the algae. He said okay, so we purchased a good portion of what he had already harvested." Daryl's personal algae supply was secure once again.

"Ray and the others started eating the algae. We gave talks to groups of friends, started selling algae and sharing in the profits. We bought capsules in bulk and filled them with algae at home, did our own labels and literature.

Business was pretty good, and the next spring Joe and I and some of his friends from high school went up to Oregon to harvest some more. Vic had harvested the previous year, so his barge and other equipment were still in place.

"Joe and I arrived in town and rented a house for the whole group. We bought a TV and a sofa, and that was about it." (They were moving up from bare foam rubber, at least.) "It was my first time living with a bunch of boys—what a team experience! And I've been on a lot of teams. They were all great kids and we had an outrageous summer. Finally I was living that professional parent role I'd been dreaming about.

"The schedule that summer went something like this," Daryl says. "Mike woke up quick and easy so I'd get him first. Our alarm was set for 5 AM—he and I would get ready to go, fix breakfast, and then as we were leaving, wake up the rest of the boys. They'd pull in at the lake at 5:55 AM, and we'd all go out to the barge and start harvesting. By that time we'd have the tape recorder set up so we could play music. We'd pull the algae up out of the water on a conveyor belt and scrape it off into Coleman coolers. I'd take the coolers to shore in a small boat, put them in the back of the van we were borrowing, then run it into town to Klamath Cold Storage.

"Once I got to the freezer, I'd scoop the algae out of the coolers with a grain scoop, put it on cookie sheets and set the sheets inside to freeze. I think we had about a thousand cookie sheets at that time. Later in the day, I'd take one of the boys down with me. He'd start taking the frozen algae off the sheets and put it in boxes; that job usually took four or five hours.

"We were after the maximum amount of algae we could harvest during a very brief period. So it was physically demanding, but it was also invigorating work. We were on the edge all the time. At the end of the day, the kids would start the clean-up. I put Joe in charge of that (his first opportunity to be a boss). The kids were never in a hurry to leave. The clean-up process took time and patience, so I'd go home ahead of them to start dinner.

"When they got in around 7 PM I'd have the food ready; after dinner we'd play basketball at the junior high nearby till we couldn't see any more. On Sundays we'd play basketball all day.

"It was interesting," says Daryl, "that what each of the kids did that summer was to adopt the best characteristics of the others. They had to stand on the barge in an area the size of a table 12 hours a day, six days a week, for almost three months—but they never tired. It totally changed my ideas about teams and the way people can work together. By the time they went back to school, all of them wanted to be top students like Mike, and all wanted to make first team in football like Joe. For the next three years they were hardly out of each others' sight.

"We harvested in July and August, then the boys returned to San Diego for school. I stayed in Klamath Falls and hired a new crew. The algae started to get thicker and the work got faster; it was an incredible two months. At the end of harvest, when I went back to California, Victor had decided to move his business from New Mexico to San Diego where I was."

That was really quite impressive. In the space of a year and a half, Daryl had gone from being completely broke, answering a newspaper ad for a job he wasn't qualified to do, to convincing a good-sized company to convert from solar and go into the algae business. And to top it off, his brother was now joining him as well. No wonder Daryl calls himself a progress fiend!

He was also about to experience a major revelation in personal growth. "There was a woman working at the solar company at that time," he says, "who was always having personal problems—she was a personal relationship junkie who read every self-help book she could get hold of. Her own life was a mess, but she had all the theories down and she was eager to share what she knew.

"One day," Daryl explains, "she showed me a card with a sentence printed on it. 'Here, read this,' she said—it was some nonsensical thing. 'Read it aloud,' she said. So I read it out loud. She said, 'No, read it again.' I did—and again she said to give it another try, that I'd made a mistake.

"Now you have to realize," says Daryl, "at that point in my life I believed that if I really, really concentrated, I didn't make mistakes. The result of that in Marta's and my relationship was that I always stuck to my view no matter what. So when she disagreed with me, I considered her wrong. The concept of being right was very important to me, so important that it interfered with our communication in a massive way and caused a lot of disharmony. That was one of the things that hadn't worked in our marriage.

"Anyway, I read the card a third time, and the woman again shook her head! 'That's not right.' So I got serious about doing my utmost—I summoned all of my powers of thought and focused everything I had on it. Then I saw it—I'd skipped over one word that was repeated; it appeared at the end of one line and again at the beginning of the next.

"The truth of what that meant was like a blinding flash that went through my whole being. I thought 'Holy Smokes, even when I do my absolute best, focus everything I have on something, I can be *wrong!*' It was an inner rush of understanding that changed every relationship I've had since. I realized that all I had to do was consider everyone else's opinion at every stage of everything. I thought, *I have to tell Marta.*'"

That was in November of '81. Daryl called Marta, and she understood; she heard that it was a life-changing experience for him. They met as soon as possible. "We were still madly in love and had a beautiful weekend. After that we were on the phone every night for three months, talking about how we could get back together."

Four years after their separation, Daryl and Marta were remarried, on March 19, 1982. Marta and the kids then moved out to be with Daryl and Joe in San Diego.

CHAPTER 9

New Beginnings

WITH THE FAMILY REUNITED, DARYL continued to work both sides of the newly combined solar-and-algae business, making a little money and saving a little, too. After enrolling her boys in the San Diego school system, Marta jumped in and started helping. They bought algae-processing equipment—Daryl was doing lab work now as well as learning to install and operate a freeze-dryer. They were also driving up to Oregon on a regular basis and bringing the algae back themselves in an insulated pick-up truck. So Daryl was back on a learning curve with algae and sales—when trouble struck.

"We were planning to harvest again in the summer of '82 and the boys wanted to help. Joe had finished his junior year in high school. Justin was still too young, but Sevin was going to work, so we all moved to Klamath Falls ahead of the crew to get everything ready. What I didn't know was that Vic had made a decision—without telling us about it—not to harvest that year. He had enough algae and didn't want any more. We were expecting to start right after July 4th—Joe's friends were on their way up that weekend to complete the crew. Then Victor arrived with his announcement and completely changed the face of what was to happen that summer.

I felt the way he'd set it up was unfair to all of us—to me, to Marta, to our kids, and to Joe's friends as well."

The situation created a major schism between Daryl and his brother Vic. "All I wanted was algae. You see, to me, *the amount of algae in the freezer and the opportunity to change things in the world amounted to the same thing.* Without algae I couldn't help people in the same way. The more algae I had, the more opportunity I could offer to others—it was as simple as that. We decided we'd have to call a halt to everything and not work with Victor again."

At that point Daryl's brother had all the harvested algae in his possession. He also had the lease on the lake. Daryl and Marta had enough for themselves, but not enough for any kind of future. So from Daryl's point of view, his brother was forcing him into a choice between Marta and Vic, between progress and no progress. "I really cared for Vic, but I also knew we were at an impasse. It was tough. Victor was living in our house in San Diego at the time. So we drove back down to California and when we got to the house, we asked Victor to leave. Vic stopped selling us algae, and we severed all ties. But we refused to give up on the algae business.

"I'd met some people in L.A. earlier that year who were working with algae, people with some capital. So I talked with them and we formed a corporation called Cell Tech. We figured that if *Aphanizomenon* flourished in Klamath Lake, there had to be similar conditions in some other lake in the Northwest."

All their hopes rested on that supposition being true. So Daryl began the search. He got a list of all the state universities in Washington, Oregon, and California. Then he called the head of each biology department and asked who knew the most about indigenous algae in the area. Finally, one of them produced a lead. He traced the *Aphanizomenon flos-aquae* to Moses Lake, Washington.

Daryl caught a plane and flew into Tacoma. Mt. Saint Helens had erupted just four months before, and the whole area looked as though it were covered with snow; there was white ash about two inches thick around Moses Lake. Daryl found the algae; it was *Aphanizomenon flos-aquae*, all right, but it wasn't nearly as thick as in Klamath Lake, and the water was polluted. "I've been looking at water all my life, and I know when something's dying."

Daryl rented a car and made a couple of other stops in Washington, then came zig-zagging south, inspecting every lake on the map. But he didn't find *Aphanizomenon*, or very little. Then he followed the Klamath River all the way to the ocean, looking for any spot where it slows down. But no place was very good, and again, there was pollution in the water. He headed back toward Klamath Falls.

"I'd spent several thousand dollars on the trip and I felt pretty discouraged. As I approached town, I got off the highway—I didn't know what I was doing, really. I was on Route 39 just outside Klamath Falls where the irrigation flume crosses the highway. I pulled into the parking lot, not knowing where to go. I felt disoriented. Marta and the kids were back in San Diego, and I was alone. There didn't seem to be any opportunity anywhere else, so here I was back where I'd started, at Klamath Lake. And that had been a dead end, too."

In recent months Daryl had tried to find loopholes in his brother's lease that would let Cell Tech harvest as well—for there was a virtually unlimited supply of algae in the lake. There would have been plenty for both. But it was no go. He'd also tried to get a lease on Lake Ewauna, south of Klamath—he'd found a spot where the algae could be harvested before any runoff from town. But he was told, much to his surprise, that Lake Ewauna was also included in his brother's lease. Inexplicably, the state Attorney General had pretty much closed the door to any harvesting other than Victor's. (Later, the whole concept of exclusive harvest rights was declared invalid.)

"So I was sitting there at the State Highway Department parking lot where they store gravel, feeling really down. I looked at the irrigation flume above me and then asked myself, 'I wonder what's in there?' I got out of my car and waded through a field of thistles, getting them all over my socks. Then I looked for a way to climb up—there was a steel staircase attached to the concrete. I climbed up and looked in.

"The flume was absolutely full of the most beautiful bright green algae and a huge amount of water flowing really fast. I thought, '*Oh, my God.*' I knew this spot was outside the lease system. I could feel the excitement. I said to myself, '*I wonder who owns this?*' The Klamath Irrigation District had an office just up the road, so I drove right over there and asked, "Who owns the irrigation flume? Can I talk to that person?' The guy said, 'It's under the jurisdiction of the Bureau of Land Management, Department of the Interior.' I explained that I wanted to harvest algae from the flume—and he said, 'Sure, just get an engineer to draw up what you want and I can get you a lease on the flume.'

"My heart was beating a million miles a minute, I had visions of the future, my family, everything turning out okay, but I couldn't afford to pay an engineer; so I asked, 'Can I do a rough drawing myself and get an engineer's drawings after we get the approval?' And the guy said, 'That'd be fine.'

"I sat in a hotel room for three days and drew what I wanted to the best of my ability. Then I took it down to a place on Main Street and had it transferred onto blueprints to make it look professional. I presented it with a letter of intent. Two weeks later I was granted the lease. Can you imagine the government doing a thing like that in two weeks? Miraculous!"

Of course, by now they were running out of harvest time for that year, and soon there was a frost that killed all the algae. The beginning of the new year saw the family back home in San Diego with a lease on the flume in Oregon and no way to use it. They had no money to move, no money to pay the next month's rent. Then one day Daryl got a call from a professor at the University of Nevada. He'd read a paper

Daryl had co-authored years before on the growing of algae. Somehow he'd located Daryl, and wanted to know if he would give him some ideas on how to grow spirulina in Nevada.

"I knew how to grow spirulina; I was one of the few people in the U.S. who did. But I said, 'I've got all the algae you'd ever want, and it's growing wild.' He said, 'Really?' I told him about the lake, and I said, 'All I need is the capital to develop it.' The professor said, 'I know a couple of guys with the money and I'll have them call you.'"

Daryl was just hanging up the phone when Marta got home from, as she puts it, playing demolition derby in the parking lot of a busy downtown supermarket—after trying to stretch an impossibly small amount of money to feed five hungry people. Daryl was sitting there, grinning that Cheshire-cat grin of his. "What was that about?" Marta asked.

He said, "I'm sure this will be the person who sets us up in business."

"I suppose it was just my frustration," Marta remembers, "because I totally lost it and started yelling at Daryl. I asked him if he'd given away any information about growing algae. I said, 'Will you stop *giving away* information!? You can hire yourself out as a consultant, but you should not give away any more information! *We can't even afford to move!* That requires a truck, and we don't have enough money to do that. You have to be realistic—your ideas have to be marketed, not given away!'" She laughs. "I never expected to hear from those people again."

But sure enough, a few days later, Daryl got a call from the potential investors; he agreed to meet with them in Los Angeles. Since Daryl didn't have much marketing experience, he lined up someone to take along who had sold products through network marketing. They went to L.A. and Daryl made the presentation. The two Nevada investors, whose names were Don and Neil, were impressed.

They were all for it. But there was a catch. And it was a shocker.

These men didn't want to invest in Cell Tech—they wanted to own it! They said they'd give Daryl a job managing the operation. There'd be no job for Marta. "That was their deal. They'd pay us the value of the company out of profits later, if there were any profits." The way Daryl described it to me, it seemed something like giving your kid up for adoption but getting visiting rights.

Daryl and Marta confronted the issue, and decided they had nowhere else to go. No choice. But it was a tough one. "It almost tore my heart out," Daryl recalls. "I've done a lot of other hard things for the algae, but that was the hardest." Within a few weeks, the Kollmans had sold Cell Tech outright, along with the lease on the C canal—for no money down.

Characteristically, it didn't bother Marta to sell the company, because, as she says, "We were moving forward. At that point we'd run totally out of money, Daryl was giving away ideas and I'd taken every Civil Service Exam they had available in San Diego—so I was thrilled to death. Daryl and I would practically have sold our souls, let alone Cell Tech, to get back into the algae business. And I never once let go of the certainty that one day, somehow, we'd be able to buy Cell Tech back."

Marta said she liked the new owners and thought they had good ideas. "One of the things I admired about them was that we could always get hold of them. And they almost never said no to us. They had a quadruple A credit rating which they established for us in Klamath Falls, and whenever we needed something, they never even had to think about it. If we'd been with a bank we would have been out of business the first year. But Don and Neil had started so many businesses that they realized there's a period when you don't make a profit—so they were willing to hang in there and say, 'Okay, let's give this a try, instead of, 'No, this is never going to work'."

Daryl was told he would be in charge of harvesting and processing the algae. Don and Neil signed a separate one-year contract with the marketing guy Daryl had brought in. Modifying the plans he'd drawn up in the hotel room, Daryl went up to Klamath Falls and contracted for a very modest harvest facility to be constructed by a local builder. A tank 18 feet deep was sunk into the ground next to the irrigation flume. Water rich with algae would flow into the tank from the flume. From there it would be sprayed onto a round screen like a lazy Susan. Then the algae would be scraped off, and conveyor belts would run it onto two long vacuum tables to remove the water. Daryl also put up a small building next to the tank for processing. His concept was to remove moisture from the algae through a gentle heat process before freeze-drying it.

Marta plunged right into the work of processing, working on a volunteer basis. At the end of the conveyor belts, dry algae came off the screen in hard chips. It became her job to grind those chips into a fine powder. Using a small wheat grinder, she'd fill it with chips, punch it down, grind it, pour it into a measuring cup, then dump it into a barrel. That was the sum total of production—Cell Tech's entire output was dictated by the speed of Marta's wheat grinder. She wore a mask on her nose and would come out of the building, Daryl remembers, looking like a raccoon, with green all over her face.

The drying system proved slow and cumbersome. They were harvesting too much algae and couldn't keep up with it; they realized they'd have to freeze it instead. So they turned the drying room into a freezer. They froze the algae on site, then stored the frozen product at Klamath Cold Storage before it was shipped off to be freeze-dried.

The additional space gave them room for rudimentary offices. They used 4x8 plywood sheets for desktops, and a small square hole in the wall served as Daryl and Marta's intercom. Marta answered the phone, took all the orders, filled out the Visa slips and called them in. Justin was in grade school at the time, so Marta would go home to be with him in the afternoon. Later in the day she would pack the algae, and then drive the packages in the back of her car to UPS. The hours were long, and she did a lot of driving. But it was beautiful farmland, with Stukel Mountain almost on their doorstep and a stunning view of Mount Shasta as well.

"I always used to love driving out there, even during harvest season when we went almost entirely without sleep. Going past those farms, the cattle on the left, the sheep on the right, was always so peaceful. In the middle of winter I'd be driving along, there'd be snow on the ground, and then I'd see steam coming up—there was a brand new baby that had just been born, dropped into the snow. I was always so shocked; I found out later that if calves are born in the winter, they're sturdier. And the baby lambs in the spring were so beautiful. I miss the animals."

Finally Marta got herself onto the Cell Tech payroll, with a $5-an-hour stint making a liquid concentrate of the algae—but it didn't last long. Joe came home from college and took over the job because he needed the money for school. Once again, Marta was working for free.

Eventually Marta was showing up several times a week with packages at UPS. Finally the woman there suggested, "Why don't you let us come pick those up instead of you coming down here?" *They'd hit the big time!* But still there were lots of days when the UPS truck would pull in and Marta would have nothing to give them. "Tell you what," she said to the driver one day. "I have these aluminum cookie sheets and a nail right here. I'll hang out my cookie sheet on the days I have orders to be picked up." If the cookie sheet was up he'd pull in, and if it wasn't he'd drive on past. Days when the UPS truck went on by, Daryl and Marta couldn't help but realize how fragile their new business was.

In the spring of '83, it became clear to Neil and Don that the man in charge of marketing wasn't doing enough business to keep them afloat. So he was out. Without a new marketing plan there was real danger of the company going under. Daryl and Marta began to look for alternative ways to market the algae, but concluded that despite their initial unhappy experience with network marketing, it was still the best way to go, for three reasons. It would give new consumers a source of one-on-one support as they experienced the changes the algae brought; it would give people a business opportunity that was out of the system; and it would start building a far-flung network for Daryl's dream of helping kids.

Neil said, "Okay, if you want to do network marketing, go ahead. We'll give it one more try—and keep the business going a while longer. But your salary's cut in half, Daryl."

Chapter 10

Breaking Through

After studying several network marketing companies, Daryl and Marta drew up a marketing plan. At that point Daryl decided to learn first-hand about how network marketing was done in the field. He set up a meeting in Portland for potential distributors and prepared an intellectual treatise on the algae. Then he threw his back out, drove to Portland in considerable pain, and almost bombed completely. He tried out his talk on Marta and Joe beforehand and they both almost fell asleep. "We *begged* him not to give that speech."

But Daryl stuck to his guns, and got up in front of a group of people who, as he recalls, all seemed to be wearing Spandex and polyester. Fifteen minutes into his talk he was getting nothing but blank hostile stares. He started to panic, then finally broke down and began to relate his own experience with the algae. The audience visibly relaxed, and when he finished, they actually applauded.

At a second meeting the next day only the woman who'd organized it and her two kids showed up. Daryl, who'd imagined himself lecturing to a fair-sized crowd, remembers realizing there was no reason to stand up. "So I just sat down and talked about my experience with algae again. Again it worked."

He began traveling and giving talks. Sales increased, and Daryl decided if he was going to be making money for other people, they ought to be people he liked. He first thought of Jim Carpenter, whom he'd worked with in Lumberton, and Jim's wife Stephanie. Also his friend Ray, the sales manager who had hired him at the solar operation in San Diego. He flew to visit them and signed them up. Then he went back to Hilton Head and talked to Chris, a schoolteacher who'd come out to New Mexico one summer to work with Daryl on the algae. He recruited Chris as well. All of these people began to build active networks.

So by 1985, three years from the time that Cell Tech was founded, it was at last beginning to be profitable. Marta was earning a small salary, but Daryl was still working at half his former pay. As orders picked up, Marta recruited a woman she knew from the local printshop. Shireen would become her helpmate and fellow workaholic for the next five and a half years. "Until others joined us a couple of years later and pointed it out," Marta says, "it never occurred to either one of us to stop for lunch." Shireen was someone both Daryl and Marta trusted, so occasionally Marta was now able to join Daryl on trips.

During harvest, which ran 24 hours a day for 120 days, a truck would leave every two hours to transport potato boxes filled with ziplock baggies full of freshly harvested algae. They were taken to Klamath Cold Storage, where Marta would work in the frigid environment for hours at a time. "I'd go into the corner of this huge freezer floor. The guys would carry in the boxes—they were really heavy. I'd set out the baggies to freeze. I'd be on my hands and knees—I'd reach back, take a baggie out of the box, lean forward, place it out on the floor, then grab another and just keep working my way backwards. By the time I finished, it looked like I'd covered the floor with green patterned linoleum. It was fun putting down hundreds of gorgeous green baggies, little soft pillows. I'd lay them perfectly edge to edge."

As the company grew, the harvest site was developing into quite a colorful complex. When they spilled out of their original little building, Marta bought a used classroom trailer from a church, and moved the plywood desks and packing tables into it. To one side of the trailer were two redwood-and-plexiglass double pyramids—each consisting of a pyramid resting on a mirror image of itself built down into the ground, where the cool earth could keep the inside temperature constant. Within the unusual structures, several varieties of sprouts were being grown for a new Cell Tech product. Daryl had come up with the idea to mix rock dust and water from the lake and grow buckwheat, alfalfa, and sunflower sprouts, then freeze-dry them along with some algae as toppings for salads, potatoes and such. Neil was at first confounded by the idea, but eventually he decided to go along with it. Daryl's enthusiasm was infectious.

"When the inspector from the Department of Agriculture came," remembers Marta, "he was astonished that there were absolutely no bugs and no mold in the pyramids. He told us that most other sprouts operations were plagued with mold and slime, and had to resort to chemical rinses." Was it the magic of the pyramid shape? The inspector was scratching his head when he left. Then, in some happenstance display of whimsy that was uncharacteristic of bureaucracy, the State Highway Department came along and created a statement of its own with a huge load of gravel that formed a pyramid which matched the other two. Together all three pyramids marked the little Cell Tech outpost near the foot of Stukel Mountain.

When it came out on the market, the sprouts salad topping proved so popular that it was impossible to keep up with the demand, and for that reason the product eventually had to be discontinued. But in the meantime it made for a wonderful ambiance at the harvest site. Marta kept paper plates and a jar of salad dressing handy so anyone who got hungry could go out and graze in the pyramids or in the huge garden where giant vegetables were grown in algae-enriched soil. Shireen would go out in the middle of the day and sing in the pyramids, and Daniel, a French organic farmer who'd been brought in to help set up the sprouts operation, would work in the garden, planting his magic seeds.

Daniel was like a dynamo, a petit man in a huge wide-brimmed hat who was always moving, always working. He was anguished when sheep from the fields nearby would come and chew up his garden. Complaints did no good. As soon as he saw the sheep, he'd be beside himself. He told the farmer again and again, and the farmer tried, but apparently it's hard to stop sheep if they want to get somewhere. Daniel never knew about the final victory, Marta says.

"After he had left for Minnesota to another job," she recalls, "I was alone one night at the harvest site. I was almost always there late, especially at the end of the month when I was figuring up commissions. Nights when there was no moon and no stars it was so black out there, sometimes I couldn't find my car when it came time to leave. One dark night there was a knock, and I went to the door and opened it. It was the farmer from across the way. 'Here, you deserve these,' he said, and handed me a paper bag. I looked in the bag—it was full of lamb chops. I didn't know what to say."

For a long time Marta and Shireen continued to do all the bookkeeping by hand, driving or sending it up to Washington to be entered on a computer. Then they acquired an off-brand database unit just for keeping names and addresses. Marta knew little about how to operate computers, and was often reduced to tears when she was having problems with it.

On one of her searches for new software she was referred to John Wilson of The Brainstorm Group in Ashland. He and Marta hit it off on the phone, "probably because we were both from the Midwest," she says. "One of the first questions John asked me was, 'How often do you back your system up?' I paused, then I said, 'Back up?' And he said, 'You mean in two years you've never backed up your work?' And I said, 'I'm not even sure what that is. Could you explain it?' By then he was laughing so hard he could hardly talk. So I put the computer in the car and drove to Ashland. John was extremely helpful."

By 1987 the company was finally outgrowing its old site. On her trips to town Marta began looking for a new home for Cell Tech. She had always loved the old auto dealership building on Main Street—the huge windows, the columns inside, those high ceilings she is so partial to—and like their double pyramids, the old Balsiger building even had an Egyptian theme. Driving past one day, she saw that it was empty and had a real estate sign on it. She phoned right away but was disappointed; they couldn't afford the rent. After consulting Don and Neil, she called back and said, "How about if we offer you half? Then we'll increase that every year for five years as we get more successful." Her offer was accepted.

When they first moved into the building, it seemed vast—it was almost impossible to find anyone. But before long it began to fill up. Joe came home after a couple of years on the road with his band, and began to help his father design production facilities. In the great "outback," the region that had been a storage area for cars, Joe constructed an entire building-within-a-building to ensure that the bottling plant would remain free of dust and other contaminants. Bit by bit, an encapsulating and bottling line was installed. Out front more operators were added to take the growing number of phone orders, and a shipping department began to take shape. Offices were built for an accounting department, and for Don and Neil as well. In another part of town, Cell Tech acquired and began to operate its own freeze-dryers.

As the operation grew, it became clear that the C canal where they were harvesting was limited in its capacity. Regulations on the irrigation system restricted the amount of water they could take out or put back in; they weren't permitted to create turbulence. So the size of the company would be forever constrained by the capacity of the canal. Daryl investigated and found that another canal was available which had a lot more possibilities. For one thing, there was a 20-foot drop in elevation just before the potential harvest site which would aerate the water, adding carbon dioxide, oxygen, and sunlight, thus promoting new cell division just before harvest. Daryl said to Neil, "We can get that location." Neil said, "Let me see a drawing."

Marta and Daryl had the drawing prepared and wrote to Neil, explaining that after the initial investment, harvesting would be less expensive. But Neil said no. "We're satisfied with the amount of algae we're getting out of the C canal." The company was starting to make good money at that point, but the investors were in the algae for profit rather than for what it would do for people. They didn't want to reinvest to help the company grow to another level. They didn't share Daryl and Marta's dream.

Now more determined than ever to repurchase the company, Daryl and Marta noted on their calendar the day Neil had said no to the A canal. Then they scraped together their savings and leased it themselves. "When we announced to them we wanted to buy the company back," Marta recalls, "they named a price that we felt was ludicrous. There were countless offers and counter-offers, but most of them seemed to have a fatal hook in them. It was a grueling, horrible year. I remember once they said they'd sell it to us for a huge amount of money, and in addition the investors and their children would continue to be paid a percentage for the rest of *our* lives. It took almost one full year to buy Cell Tech back.

"Those were tough days. Daryl and I were supposed to be inspiring distributors, traveling and training, but we were doing it for a company that I thought we wouldn't even be working for pretty soon. I said to Daryl, 'This time next year I'll probably be somebody's secretary and you'll be teaching school.' We'd still try to be enthusiastic for the sake of the distributors, and they'd say, 'What about this or that next year?' I wanted to say, 'There probably won't even *be* a next year,' but instead I'd say, 'It's going to be great! What a year!'"

At last all of the Kollmans' efforts were successful. "It's been three years since we finally bought the company back," Marta says. "We'll be paying for it for ten, but then it's over! At first, when the price was really high, I said to Don and Neil, 'Here's my suggestion: You loan us the money, and we'll pay you back.'

"They laughed at me. 'Why should we?' they asked. I said, 'I know I'm not the business person you are, but what's wrong with that?' They said, 'Why should we loan you millions of dollars for a company that's not worth it?' Of course *they* were the ones who'd set the price!"

In the end the investors saw it Marta's way, probably realizing that without its extraordinary founders, Cell Tech's value would be greatly diminished. In August of 1990, Daryl and Marta re-purchased Cell Tech, and soon after, they installed a new harvest facility that was light years ahead of the old C canal. It was 15 years since they had begun their quest.

For Daryl and Marta, perhaps the hardest part was over. They'd secured a substantial economic base from which to build the global network they'd dreamed of. Each year millions of pounds of algae would be harvested from Klamath Lake and distributed to people all over North America, and later the world. That job would be largely in the hands of the distributors, who by now had already quietly built a multi-million dollar network.

Who were these distributors and what were their stories? I wanted to know. Daryl says that "algae is a magnet that draws caring people together," but I wondered if the top distributors really saw the algae as an avenue for global change—or just as a means to a private fortune. Were they motivated by material or humanitarian desires? Or by a healthy combination of both?

To find out, I decided to trace one highly successful branch of the Cell Tech family tree, six distributors in a vertical line who achieved Cell Tech's highest rank—Double Diamond. I wanted to see how it had begun and where it had led—how passing the algae from one hand to another had altered lives. And how the money which came to them as a result had affected each of them—I knew there was a good deal of money involved.

In the course of my talks with these people, I was struck by several things: 1) how often it's those "insignificant" human contacts that make the crucial difference in our lives, 2) that when opportunity knocks, most of us don't hear a thing! 3) that people can succeed in network marketing either by accident or by design—although it's a whole lot surer if you *work* your business, 4) that the variety of people who can do well at network marketing is almost infinite.

I also learned, despite my own resistance, that network marketing was not the greedy "pyramid scheme" I'd always believed it was—not by a long shot.

Part III

ABUNDANT OPPORTUNITY

Chapter 11
So Show Me, Showshawme

DARYL SAYS THAT NETWORK MARKETING is about redistributing wealth, and I believe it. A young African American woman at one of Daryl's talks called it, "the most humane form of capitalism," and that's a pretty apt description, too. That it levels the playing field for some of the downtrodden and disadvantaged of this world was never better illustrated than in the story of a man named Showshawme.

In the beginning, Showshawme was a perfect example of the kind of child Daryl Kollman was trying to help—he had everything in the world going against him. Today, Showshawme is responsible for the creation of a network that produces over 90 percent of Cell Tech's whopping multi-million-dollar monthly sales volume. I was soon to learn why. Showshawme is an unusual person.

Sometime last April, having arrived at the Berkeley Marriott late at night for a Cell Tech Regional Rally, I was sleeping soundly in my room when the telephone suddenly jangled me into morning. "Wake up and celebrate the algae!" a mellifluous male voice urged. Who on earth? Then I realized. Of course, it had to be the fabled Showshawme.

We met a few hours later just as the rally was about to begin. A dapper man in a tan jacket and dark slacks, with reddish brown hair and a closely trimmed beard,

Showshawme ran up on stage to wild applause with a smile that seemed to me both proud and shy at the same time. Though he's slight in build, Cell Tech's star distributor exudes such intense energy it's impossible to avoid noticing him even in a crowd—and there was a crowd of Double Diamonds on that stage, all of them clearly fond of him. He's legendary for being generous, and persistent. And it's apparent when he talks that Cell Tech is his life.

Born a "blue baby" in industrial Hartford, Connecticut, in 1952, Showshawme (it's not the name he was born with, but we'll get to that in a little bit) weighed just two and a half pounds at birth and spent his first months of life hooked up to tubes in an incubator, struggling to survive. His father worked for a machine tool company, his mother at various factories and dry cleaners. Both parents were alcoholics and physically abusive.

As a child, Showshawme was mischievous and curious (actually, the description is still appropriate). "I always wanted to know how things worked, what made them tick. At the age of five I got into a yellowjackets' nest, was stung hundreds of times and almost died. It was a very traumatic experience, and I'm sure it did something to my psyche," Showshawme told me. "After that I began getting into a lot of trouble. I remember being jealous of my sister because she could stay up later—of course, she was seven years older than me."

At first Showshawme was the best reader in his class, then he became the prankster, the "instigator." The only subject he liked was Show and Tell. "I got suspended from school a lot—for throwing spitballs, knocking over furniture and setting off firecrackers. Once I turned on the school fire alarm when it was pouring rain and made everyone go outside. I got caught because I was the only one inside and the only one who was dry.

"I was hyperactive," Showshawme goes on. "I ate the average American diet—total sugar and junk, heavy on the meat and dairy, but I didn't know any better and my mother didn't either. I couldn't sit still for a second. I wanted to get more attention, I guess—or that's what they told my parents, anyway."

Though he was very bright, Showshawme's report card was mostly D's and F's, except once in seventh grade when he got an A+ in history. "I don't know exactly why; I think it had to do with the teacher. She saw some potential in me—and there was a kid in the class I was kind of competing with."

His only other achievement was to become one of the *Hartford Courant's* top paperboys—Showshawme says it was the goal-setting, the chance to win all those prizes and trips the newspaper offered that drove him to succeed. And succeed he did, winning a number of the contests hands down; he was already showing his drive and ability. But he was small and he had an attitude. He'd beat up smaller kids, and then in return he'd get beaten up and have rocks thrown at him by the bigger boys. "If it hadn't been for my older sister, who has always been helpful and kind to me, I don't know what I would have done."

By the age of 14, Showshawme had followed in his parents' footsteps and was becoming alcoholic. He also sniffed glue and took his mother's diet pills and tranquilizers. Then he added pot and LSD to the mix. (A wonder he didn't die from some of the combinations he put together—but he seemed indestructible.) The Bob Dylan song *Everybody Must Get Stoned* expressed his mission in life at that time. "I wanted to get the whole world stoned."

Predictably, Showshawme became a high school dropout. He got a job working at a Shakey Pizza Parlor where they made 28 different kinds of pizza, and became the fastest at making them. "I had the keys to the whole place. I was the janitor, bottlewasher, prep cook and salesperson. Whenever they had a new product, like fried chicken, I'd sell more than anybody. I almost became a manager; I could have, but I was too much into drugs."

Arrested twice on drug charges, Showshawme volunteered his way into a rehabilitation program. There he had 13 months to question the way his life had turned out. "I spent a lot of time reading while everyone else was playing cards or trying to think of ways they could get drugs smuggled in and get high. I got my high school equivalency diploma while I was in there and I read a lot of books about changing your life. I think the big change happened when I read a book called *Wisdom of the Mystic Masters* that was about 30 years of Rosicrucian study condensed into one book. I'd been brought up Catholic; my parents didn't go to church much, but they forced me to go a lot.

"After I got out of rehab I kept on reading about different philosophies and religions around the world. I moved back into my parents' house and worked in a factory doing a robot job for nine and a half hours a day at minimum wage—I did that for two years. Finally I quit the factory and took a live-in job at a hospital for crippled children.

"I worked there for two years as a porter, taking patients back and forth in wheelchairs or in their beds to appointments, wheeling them all over the hospital. Also I cleaned out the incinerator twice a week, scrubbed toilets and waxed floors. I hated those jobs, but I didn't know what else to do. You weren't allowed to have interaction with the kids. I tried to do that and got disciplined for it. I had real long hair at the time, I was a rebel. I wanted to change the world, to make a positive difference—I was actually idealistic—but I didn't know what I was doing."

Showshawme was becoming increasingly bitter and frustrated as time went on. He could find no way to channel his idealism, and he had no one to help or advise him. He remembers the violent thoughts he sometimes had. But then something happened that changed the path he was on: He became interested in health food.

"It had started a few years before when I was trying to find a cure for my acne. Gradually I became more interested, and before long I was the only employee who wouldn't eat in the hospital cafeteria. I'd go back to the staff house where I lived on the grounds and have some fruit or carrot juice or yogurt or sprouts.

"In 1973 I became a total vegetarian, and except for a little butter on something once in a while, that's basically continued till the present time." Showshawme feels the change in diet helped him change his attitude. "Till then I'd blamed my parents and society for everything that had happened to me; now I realized I had the power to create change."

In 1975 Showshawme read *Survival Into the 21st Century* by Viktoras Kulvinskas, a raw foods advocate. Years later Kulvinskas would become one of Showshawme's top Double Diamonds, but at that time all Showshawme wanted to do was tell him how impressed he was with the book. The two met, and Showshawme then began promoting Kulvinskas' work on his own. "I had brochures and flyers made; I got him on some television and radio shows. I had no experience doing it—I just learned it by myself."

Showshawme at that time was experimenting with both raw foods and with macrobiotics, which strongly emphasizes cooked foods. In the mid-seventies, as the health food wave was beginning, these were two of the major schools of thought—and there was tremendous controversy between them. Showshawme felt both had their merits.

He began working health shows and conventions with Viktoras, selling sprouts, making juices, and marketing Viktoras' books. He also invited teachers from various disciplines to come and give talks in his apartment and in rented halls. "I'd travel to health shows and people would say, 'You're from Hartford? How come there's nothing happening in the health field there?'—so I'd go home and try to make something happen."

In 1980, Showshawme decided it was time to make a complete break with his past. By now he was drug-free and feeling better about himself, but he longed to live in a place where there was clean air and water, warmer weather, and abundant organic fresh fruits and vegetables. *Hawaii*, he thought. That was the place to be. He couldn't afford to move, but he wasn't about to let that get in the way. He talked a couple he knew into going there first—hooked them up with somebody he'd met from Hawaii who was visiting Hartford. Four months later his friends bought some land on the big island and invited him out. Success! Hawaii it would be. Showshawme flew out of Hartford on a one-way ticket with $200 in his pocket.

"I changed my name legally about seven days before I left. I wanted to have a whole new identity. The name came to me kind of intuitively; I don't know exactly what it means. Some people say I could mean it in a joking way. You could say So-show-me; pronounce it that way and it's like Show and Tell, the only thing I excelled at in school."

Perhaps Showshawme had imagined a life in Hawaii of health, harmony and sweet hibiscus, but what he encountered was something quite different. When he first arrived, he stayed at his friends' place for a time, then decided to strike out on his own. "I had a little pup tent (but no experience in camping), and I went out to live in the jungle. But it rained a lot and all my belongings got mildew. I was on food stamps; there were millions of mosquitoes. I had no electricity and no running water. I had to catch rainwater and deal with rats eating through my tent.

"To support myself, I started a sprouts business with a girlfriend. I hitchhiked from the jungle to town about eight miles away and grew sprouts in her open-air garage. Once a week we'd go deliver them in her car. Then I heard about spirulina and started eating it. I liked the concept about feeding the world and that it was a vegetable protein. So I became a spirulina distributor."

Around 1980 someone told Showshawme about the algae from Klamath Lake, but he didn't pay much attention. He thought it couldn't be any better than what he had. "The guy who told me about it didn't have any literature, so I just scoffed it off." A few months went by and he ran into the fellow again. Reminded, Showshawme sent away to New Mexico for literature.

By this time he had moved out of his leaky pup tent into a 5x8 foot storage shed at a 2,000-foot elevation in the Hawaiian rainforest. In exchange for free rent, he stayed on the property as a caretaker and looked after three pitbulls and a lumber supply. He lived in the shed eight months, then bought himself a 16-foot dome tent and persuaded the landowner to build a platform for him so he'd be out of the mud. The new tent was his home for a year and a half, and it was during his "new tent" period that he first tried the algae from Klamath Lake.

"I ate four capsules. Actually, I took it out of the capsules, mixed it in some water and drank it. Within half an hour I felt as if I was going to do somersaults over the tent, I had so much energy. It was like all the cobwebs in my mind were gone. I felt a sense of purpose come through me—I just had to share this with the world." Every day for the more than ten years since, that's been Showshawme's primary mission.

His method of marketing the algae was simple. "I'd hitchhike with my backpack 16 miles into town every day. When people would pick me up, they'd ask what I did. I'd say, 'I'm into algae,' and then I'd tell them about it. I felt more energy from the new wild algae than I ever did from spirulina. More clarity, too, though I only ate a gram, compared to eating 20 to 30 grams of spirulina. By the time the people would let me out of the car, I'd usually sold them some algae. And I told them if they got other people to eat it, I could give them theirs cheaper. Or they could possibly get theirs free if they told enough people.

"So I was already networking before I knew what networking was. It was like what I used to do when I was eight or nine years old: I'd save up my allowance and make a deal with the owner of the candy store to buy a whole case of candy. Then I'd sell it to the kids in the neighborhood and have a few pieces to keep for myself for free."

By 1983 Showshawme had made enough money from his algae sales to fly to Los Angeles and attend the Whole Life Expo there. That's where he first met some Cell Tech people. He spent the next several years marketing Super Blue Green Algae.

In August of '85, Showshawme finally moved back to Connecticut. "I was basically broke. The networking was happening in Hawaii but there was no literature and the local economy was unstable. So at times I did well and at other times I didn't do so well. When I moved back, I lived at my aunt's house because my parents had passed away during the time I was gone." He pauses. "So they never got to see my success." His regret is unmistakable.

To supplement his algae income, Showshawme taught firewalking and sold macrobiotic supplies and herbs. Determined to be successful, he called Daryl Kollman (they'd never met) at Cell Tech and invited him to come to Hartford and give a talk. "I got such a sense of commitment and purpose in that call with Daryl. And it wasn't even hard to convince him to come."

After the talk, the two sat half the night in a church basement and talked about the vision they both shared. "At that time I realized I still thought that money was evil, so I started doing affirmations and reading books on prosperity." Evidently it helped, because one day not long after that, Showshawme made a routine telephone call that would eventually result in millions of dollars of business for Cell Tech, and enormous income for himself as well.

CHAPTER 12
From Network to Team

THE DAY HE MADE THAT first critical contact with Kim Bright, Showshawme (for once) wasn't even trying to sell algae. He had read an article in the *East West Journal* about quinoa (keen-wa), an ancient South American grain he was interested in, and decided to look up the import company mentioned in the article. They told him there was a distributor about an hour and a half away from him—a company called Amber Waves.

"So I picked up the phone, called the owner and ordered some of the grain. Then in the course of the conversation I told her about the Super Blue Green Algae from Cell Tech. She wasn't very open about it because it was only in capsules at the time and macrobiotic people believed anything in a capsule wasn't a food, it was a supplement. 'I don't do supplements,' she told me rather coldly. I did my best to convince her it was a whole food. 'All you have to do is dump it out of the capsule,' I said. I argued that after all, macrobiotic people were already eating algae from the ocean, like nori, kelp, wakame, hijiki. But she wasn't convinced."

It was not an auspicious start for a business alliance that was a long shot at best—for culturally, and in life experience, Showshawme and Kim Bright were worlds apart. Kim, whose family was in the restaurant and nightclub business in Denver, had been a natural leader growing up. Nurtured and loved at home, and looked up to by the other kids in her school (literally), Kim was six feet tall, with handsome features. She'd become the first female drum major in her school, had gone to college in Europe, had sung with the Berlin Philharmonic, studied with some well-known voice coaches, and attended a dance academy in Beverly Hills.

However, like Showshawme, she'd also had her own struggles with alcohol and had found an answer in health foods. "When I was out in L.A. I started going down the tubes, drinking heavily and living in the fast lane. One night I drank till I passed out. In the morning when I came to, I looked around the Palm Springs condo and realized what I was doing to myself. It was a transformation. I went cold turkey, straight off a bottle of booze a day into a 14-day fast, just vegetable broth. It's a wonder I didn't die—I had the D.T.'s for three days, then walked into a health food store for the first time. Someone told me to get some protein powder. I did, and also picked up a copy of *Prevention* magazine."

Kim started drinking raw milk, eating organic fertile eggs, and reading everything she could on nutrition. Eventually she opened a macrobiotic learning center, a whole foods distribution company, and a macrobiotic restaurant called *A Change of Seasons* in Westport, Connecticut.

"A strange character called me one day in late '85 looking for quinoa," Kim recalls. "He said his name was Showshawme. I thought he was Indian. He told me about the algae, but I wasn't interested.

"Then he started calling me up every single night for two weeks," says Kim, "and we got into amusing and strange conversations about algae. Finally I said, 'Look, this has got to end. I've heard enough and my boyfriend's getting upset.' I told him I'd agree to meet him for half an hour and look at what he had. 'If I want it, fine. Otherwise, don't call me anymore.'

"He showed up, looking kind of wild and hippyish. And I'm a little more conservative—my boyfriend was too.

Showshawme had brought a poor quality video showing someone going through a supermarket. Then Daryl came on, and all of a sudden I got very interested. I thought, '*Kollman's a visionary. He's looking at the entire world and that's the game I want to play.*' He came across as having a tremendous amount of integrity, truth, and spirit.

"I decided to try the algae. And I did, but nothing happened, so I called Showshawme and told him, 'Still nothing after 23 days.' He said, 'You're not taking enough.'" At the time Kim was taking two or three capsules of Omega and Alpha (the two basic Cell Tech products) per day. She upped it to four and felt an immediate response. "I went through some tiredness, and had some other symptoms, then came up to a new energy level, with clearer thinking. I'd never been able to do administrative work for very long at a time; I'd always have to jump up and do something else. Now I noticed I had more staying power, a steadier energy. So I thought maybe there was something to it. After all, it *was* actually a food, and had a lower sodium content than seaweed.

"I began telling clients about it when they came in for consultations at the Macro Center. Some of them were curious just seeing it on my shelf. But I told them I wouldn't sell it to them because I was still doing research on it myself. Finally one woman said, 'I insist that you sell it to me today!' That was my first sale. She got immediate results and came back, wanting some for her sister and daughter." Kim called Showshawme and told him.

"So why don't you become a distributor?" he asked.

"Look, thanks, but I've got four businesses already," she answered.

Showshawme recalls that it took him several months to get Kim into the business. "I persisted and she resisted, but I never gave up. I just knew that the macrobiotic people would like this algae. And she was a good link to them. They had to know about it. They *had* to give it a chance."

Finally Showshawme went to the Macro Center and did a slide show for thirty people. Twenty-eight of them signed up, and it started to grow from there. Word spread, and Kim finally committed herself to doing the algae business.

One of the lecturers at the Macro Center at that time was slim dark-haired Steve Gagné, whom Kim had known years before. I talked with Steve over lunch one day at the Berkeley rally. A group of us had invaded a natural foods supermarket with a sit-down deli where you could graze on about a hundred different vegetarian dishes from around the world. I felt immediately comfortable with Steve; he talks to people he's just met as though he's known them for a long time.

"Kim had been a student at the Kushi Institute in Boston back when I was teaching there," Steve told me. "I didn't know her well, but she always impressed me as creative and innovative; I remember she was making macro cookies and cakes and selling them to people. After she completed her course and got her certificate, she moved away. And I moved to Colorado, so we lost touch.

"Several years later, I ended up back in Boston and Kim was in Connecticut, where she'd formed a whole health center. She invited me to teach there. On one of my visits she told me about this blue-green algae. She said, 'I've found something that can help people and make money too. Why don't you be smart and *do it!*'"

Steve was skeptical. "I'd taught all over Europe and the U.S.," he told me. "People were always giving me natural stuff to try, but most of it was not very memorable. I told her I'd take a look at it. She said, "Why don't you try some now?'

"So I ate some, then I gave my talk that night, and I couldn't stop, I was so full of energy and new ideas. Afterwards I told Kim, 'You've got to get me some of this.'

"She said I could get it myself and explained the marketing plan. But I didn't know anything about multi-level marketing. I'd heard of Amway, but I never knew what that meant, except my impression (mistaken, I later learned) was that you go door to door selling something, and that was something I wasn't willing to do. So I had no interest."

Kim knew that Steve was having severe financial problems at the time. She respected his abilities as a teacher and counselor, but saw that he was making less money than he deserved. Steve was working long hours traveling and teaching; often he was away from home three weeks out of the month. "I always kept my mind on my goal—that I had to earn at least $3,000 every month to pay all my family's expenses. Sometimes I made it and sometimes I fell short and had to borrow money from a friend. That's the way my life was."

Although there was probably no one who was a better candidate for earning extra income through network marketing than Steve Gagné, it didn't occur to him at the time to even think about deviating from that treadmill, nose-to-the-grindstone existence he described. Perhaps it was that very sense of responsibility to his family, I thought, that blinded him to the possibilities inherent in Kim's offer. For family is very important to Steve, I realized when I heard his story.

He'd been raised in an orphanage in Maine, sent there with his sister when he was five and she was ten. Two weeks later she was taken from him and sentenced to reform school for shoplifting. Except for weekly visits from an adored older brother who was later killed in an auto accident when Steve was 13, he was alone and without family for the next ten years.

"Finally, when I was 15, my next to oldest sister adopted me, and I went to live with her. In high school I got interested in jazz—bought myself a beat-up saxophone and paid for it working in a chicken factory. When the State of Maine asked me what I'd like to do when I finished school, I told them I wanted to go to music college. They paid half since I was a ward of the state; I got a loan for the rest. After three years I moved to Boston and got a band together. We played nights and I got a day job at Erewhon Natural Foods, the first natural foods company in the country. I started to study macrobiotics and several years later became a counselor and teacher with Michio Kushi." (I could see from Steve's choices at the supermarket deli that he was still just as interested in healthy foods.)

"When I left Kim's place that week after I first tried the algae, I went on to several other cities to teach. I was on the road for a week or so, and when I got home, there were several messages from Kim on my answering machine. Because what I'd done was I'd told *everyone*, every client I'd seen, about the algae and what it had done for me. I'd told them it was a fantastic product and they really should try it; then I'd given them Kim's phone number.

"For a while I didn't return Kim's calls—I was happy to let her sell it to them; I just wanted people to have it. But she called me up a few days later and she was almost swearing on the answering machine till I finally picked it up. Like, 'You big dummy! Come on, wake up! Why don't you just sign up? You've got all these people interested, and they're calling *me*. I don't want to take them under me—they're your people—so sign up!'"

But Steve still didn't get the message. Kim says, "I would send the people some algae and tell them that Steve wasn't signed up yet but he would be soon... Eventually there were so many people calling me I began to think, *What's wrong with Steve?* Like a lot of people into macrobiotics, Steve had a poverty mentality. I was more business-minded. I didn't believe money was evil—I just thought we needed to redistribute wealth so the people who would utilize it for good would get more of it."

"Finally," Steve remembers, "she set Showshawme on me. He came up from Connecticut—I was in Northampton, Mass.—and was so insistent I thought he was totally whacked out. We had guests at the time, so I got only about half of what he was saying, and rushed him out. But he called me back, and kind of *made* me commit to signing up.

"Then I had about ten days without any work and I was hanging around with Richard France, who was a specialist in Oriental medicine, and writer John Mann. They were also in Northampton; we were all into macrobiotics, lecturing, writing, teaching courses in health.

"I guess we were all having a 'short month,' says Steve. "Anyhow, I started talking to them about the algae. Richard got excited about it right away, so I gave him a handful of capsules. He tried them, liked them, and signed up. It took John [who would later become one of Steve's Double Diamonds] about a month to get the $25 together to get started; he was totally broke at the time.

"Richard took off like a shot," Steve recalls. "He went out on the road and started distributing the algae. Pretty soon I got a check in the mail for $73 from this company called Cell Tech. And I said to myself, 'Wow, what's this? I didn't even do anything!' I just told my friend Richard about something I liked, and he did the rest."

I was soon to learn that Richard's enthusiasm for talking to people about the algae was uncharacteristic. He's a loner—so quiet and self-contained that it's hard to get much going even in an interview. The scholarly Richard France is perhaps the most unusual example of a network marketing person that you could ever hope to meet or talk to—and hoping to meet or talk to him can itself become quite a project. Merely having *had* an interview with him became a hot topic at Cell Tech—"You actually talked to Richard for *ten minutes?!* Amazing."

At our meeting in Berkeley, I found Richard to be slightly preoccupied, philosophical, introverted, definitely not conversational or chatty—hardly your typical razzle-dazzle salesman. Richard is tall with medium brown hair, a strong face and good bones. I surmised he was about 40—(of course he might be older—so many Cell Tech people look younger than they are). But I didn't want to ask; I was just grateful he answered the few questions that I did put to him. Every sentence Richard utters seems to have been weighed before he puts it out there—as though his life were measured in words rather than years, and after he's used them up, it'll all be over.

"I'd been interested in philosophy in college," he told me, "and did my Master's in world religion at Indiana U. I thought I'd probably go on to teach college, but one thing led to another and I got into Oriental medicine as a profession. Steve Gagné and I had similar interests. I'd been a teacher and counselor in Oriental medicine and macrobiotics for about ten or 12 years when Steve turned me on to the algae." Richard paused, seeming to search for the best way to explain what happened.

"I'd tried both spirulina and chlorella," he said finally, "and didn't especially like them." He paused again. "But there was something about the Super Blue Green Algae that broadened my vision. I think I was stuck as a person, and the algae helped me to continue growing—that's what happened. I felt a kind of quality from it that I'd felt only in other wild food—I was a wild foods enthusiast. So I recommended it to people in my counseling work."

Like his friend Steve, Richard didn't pay any attention to the business side of it at first. "In fact, the way I finally really got started was when someone from my downline called me and told me he was ready to make Executive, so I'd better hurry up and do it, too. The marketing plan rule was that I had to make Executive before anyone in my network did, or I'd be passed over and lose them. 'All you have to do is sign up one more person and you'll make Exec,' my friend told me. I was standing at someone else's house doing counseling when he called, so I just turned to that person and said, 'How would you like to be an algae distributor?' They said 'Fine,' and within the next ten minutes, we had it all set up."

In spite of his natural reticence, Richard's business began to grow. One of his most active builders turned out to be Loren Spector, a tightly coiled bundle of energy with dark curly hair, a chiseled face and very wide smile. Loren has an abrupt, no-nonsense style about him. "New York people think I'm a New Yorker because I talk as fast as a New Yorker, but I'm just a kid from Cleveland." He flashes those teeth. A serious tennis player from a "very Jewish, very middle class" neighborhood, Loren was shy and self-conscious till he went to EST and learned how to speak in front of groups. Eventually he went to work for EST as a Center Manager, then later became Director of the Macrobiotic Learning Center in Brookline, Massachusetts.

"About eight or ten people, many with serious health problems, would come and stay for a week at the Center, and we'd give them a crash course in macrobiotics. There were two cooking classes every day and lectures every night—we had acupuncturists come in, the whole shooting match. Richard France would come once a week to do dietary counseling.

"So one day he comes in and before the counseling session he says, 'Loren, I've got to tell you about this incredible product I'm working with.' I took one look at it, saw it was a pill, and immediately discarded it, because macrobiotics people don't take supplements, right? And when I found out it was multi-level marketing, I told Richard I *really* wasn't interested.

"Well, Richard comes in the next week and he says, 'Now Loren, I know you're not interested, but . . .' And then he starts telling me some algae stories." (*Richard telling algae stories?* I reflected that he'd been using up a lot of those precious words of his—he must have really believed in the algae.) "By now I was getting a little annoyed," Loren told me, "and I said, 'Richard, I really don't want to hear about it.'

"Three weeks later he comes back and he says, 'Listen, Loren . . .' Now Richard is a very laid back guy, and he was ruthlessly patient with me. He said, 'You know this is so important . . . I just want to have you . . .'

"'RICHARD!' I said. And by now I wasn't being nice about it. 'Shut up! I don't want to hear about this stuff!' But he was unstoppable, and it just went on like that. Finally I figured I'd better try it 'cause this guy wasn't giving up. So I said,

'Okay, you got me, you convinced me. Give me the number—I'm ordering *right this minute* so you don't have to talk to me about it next week!'

"When I did try the algae I noticed I didn't have those dips in the afternoon—I didn't take naps any more. And my tennis improved—I started noticing I could play as hard in my third set as in my first two; there was a definite increase in my stamina, and my reflexes were better. That was exciting." I could tell it was; it was easy to imagine the intense Loren Spector rushing the net and slamming the ball past his opponent.

"But probably the most important thing I noticed," he continued, "was that after taking the Super Blue Green Algae I just got totally calm. It was almost *impossible* to get me crazy! The woman I was living with noticed first because *her* life got a lot better. I was under a lot of stress—living 50 weeks a year with ten new cancer patients (many of them terminal) every week—that's stress. She mentioned to me one day that I wasn't a raving lunatic anymore, and I said, 'Yeah, you're right. I can't remember the last time I was upset.'"

So Loren began recommending the algae to his clients, and they began getting good results. Three months later he got a check for $400 from Cell Tech. He thought it was a mistake. But the checks kept coming, and after a few months he'd saved up enough money that he decided to buy a motor home and relocate to the West Coast. "We were just going to take two months off, cruise around, pick a city, and then sell the motor home.

"While we were on the road, we had our mail forwarded, and my Cell Tech checks just kept getting bigger. One day, when the latest of several came in, we said, 'Hey, let's spend another month out! Why not? We've got the money!'

"So we went up to Banff Park in Alberta—it was outrageous—and spent the whole month up there. Then when another check came in, I said, 'Hey, let's go to Vancouver.' So we had a month in Vancouver—beautiful city. And what started off as a two-month trip eventually became a two-year grand tour of the U.S. and Canada."

When he'd had enough of traveling, Loren lived in Seattle for a while, then eventually found his dream setting near San Diego where he could be at the beach and indulge his passion for tennis year round. In the years since, Loren became a successful Double Diamond. He's also become the official coach for the nearly 50 Cell Tech Distributor Empowerment Teams in North America. Using that tough-talking approach of his, he monitors the teams' bi-weekly continent-spanning conference calls, devoted to expanding both the business and the vision. Loren tells me he's living precisely the way that kid from Cleveland had always wanted to live.

Loren's sponsor Richard France is also enjoying the benefits of a sizable algae income. "These days I'm more of a Cell Tech distributor than a teacher,' he says. 'I'm having a sort of sabbatical from counseling and teaching—and becoming more involved in writing and research in areas that are gaining interest for me." Having gotten past that "stuck" place in his life, Richard tells me he's expanding his horizons in every direction.

The Cell Tech experience has helped Steve Gagné, the orphan from Maine, dispel the sense of loneliness that he struggled with for many years. "A lot of us in macrobiotics were looking for a feeling of family," he says, "and we've found that in Cell Tech. There was always a lack of love in the macrobiotic community and a lack of warmth—the people were so cold. It also lacked prosperity consciousness. That just wasn't allowed; they were always teaching us that you should be grateful for difficulties, and I never quite grasped that. I always figured I'd had enough difficulties." (I could believe that.)

"The algae also gave me a whole different lifestyle," Steve said. "I was able to be not so concerned about just making it every month. I could get the car I wanted—I didn't have to get the cheapest car I could get and then worry about a payment every month. And I was able to buy a house out here in Colorado." (For some as yet obscure sociological reason, there seems to have been a mass exodus of Cell Tech Double Diamonds to Colorado.)

Then Steve made an interesting comment. He said he credits Cell Tech with an ecumenical movement in the natural food industry, ". . . one that I would never have believed possible. There's always been a split between the raw fooders, the macrobiotic folks, and the lacto-ovo-veggies. But Cell Tech has united all these groups. You've got Showshawme and the guru of raw foods, Viktoras Kulvinskas, and we all talk to one another, seeing the benefits and pitfalls of each approach. We have one thing that unifies us—and that's the algae."

From what these distributors had told me, the algae really did seem to draw people together into community—Steve, Richard, Kim, Loren, Showshawme. I remembered what Daryl had said about the algae being a magnet that draws caring people together. It was that, yes—these were all caring people to start with. But it seemed to be more—the algae seemed to change people in such a way that they could better understand someone else's point of view. That was part of it, at least.

"Wow," I said to Steve. "Just think if they had algae at the U.N.—what that would be like."

"Right."

"Or try Congress—imagine Bob Dole eating algae. Bob Dole and Bill Clinton."

Steve laughed, then got serious again. "My sense of the algae," he said, "from all my years of taking it and witnessing the experiences of other people is that Super Blue Green Algae is the single most important food on the planet for the evolution of human consciousness. So many open-minded people constantly creating and growing, regardless of whether it's directly involved with Cell Tech or on their own."

Kim Bright says she has noticed that phenomenon, too. "There's something magical about the algae—it allows people to disconnect from the morass of confusion and meet on a common ground," she told me. "They get less worried and go forward in life. I have so many success stories that I have to believe a miracle is at work." Kim is currently living with her two children on Florida's West Coast. She's fulfilling the dream of her life, she says. She and her fiancé Ray Cassano are able to devote most of their energy and resources to their church, as well as to traveling and teaching about the algae. "Ray only started in the business a few months ago, but pretty soon he's going to be a Double Diamond too."

And what about Showshawme, the kid who used to set off fire alarms to get someone to pay attention to his pain? What does he see as the algae's most important contributions to his life? I called him up and asked him.

"When I was living in that tent in Hawaii," he answered, "I was radical and ineffective; now I'm effective. That's the difference. The money has enabled me to find out more about the algae, and to choose what I want to do with every day of my life. I don't have to be in a rat race. I don't have to drive in traffic jams—I don't like driving anyway—if I drive five miles a week it's a lot. And it's enabled me to have an indoor swimming pool with no chlorine. I can travel to exotic places, scuba dive, and," (he laughs) "ski over tall mountains at a single bound.

"But the most important part is that I finally have a family." The way he says it, you know it's true. "At first I spent a lot of time calling people at all hours of the day or night, as Cell Tech literally became my social life, my play life, my family. I didn't have a relationship at that time, so for the first few years I was totally into the algae, sharing it."

Showshawme thought for a moment. "It's fine to have the money," he said. "That's great. But what's better is that Cell Tech people are using that money for practical purposes—to heal, to help the Earth, to help the environment, and make positive changes in society."

There wasn't much more to say. The algae had brought fundamental change to all these people's lives—and their story wasn't over yet. In fact, it was just beginning.

Chapter 13
From Team to Family

In the fall of 1992, up in the Seattle area, there was a small but rather noticeable burst of Cell Tech business activity which began right after the August Celebration. And it didn't stop. Throughout the fall and into winter, one of Loren Spector's downlines just kept growing at the steady and quite startling rate of 40 percent a month. What was going on? The answer was really quite simple.

During all the years Cell Tech had been in business, its growth had been based primarily on a belief in the integrity of the product and its potential for helping the world. Money was seldom the focus. Many distributors—perhaps even most—shared the algae with others principally because they wanted to share its benefits. Now someone had come along who was recruiting active business builders into the Cell Tech family.

Rich Hosking's rationale was that it is easier to persuade a business builder to love the algae (indeed, it's practically a foregone conclusion) than it is to convince someone who loves the algae about the merits of building a business. As a result of this new outlook, there was an influx of energetic and savvy business people into the company.

The second reason for the sudden growth in Seattle was that everything Rich and his wife Donia did to build their business—all the new thinking, the new ways of doing things—they immediately taught to everyone who joined their organization. It was a step-by-step process that was easily, and some said almost effortlessly, duplicable.

A recognized hotshot in sales and sales training, Rich had begun his career at age 25, doing psychological test evaluations for the largest sales training school in North America. When he saw that the salespeople were making about three times his salary, he persuaded his boss to transfer him to sales, where he promptly set a national record in just four months. He rose phenomenally fast, progressing to district manager, zone manager, and finally becoming national sales manager at the age of 29. At 30, he was hired away by Xerox, appointed president of Xerox Learning Centers, and over the next several years was in charge of setting up 58 training centers throughout the country.

At 35 he reached a pivotal point in his life: He realized that he was unhappy. "I was making good money, I was single, I was traveling a lot—but it wasn't what I wanted. I wasn't fishing, I didn't live where there were trees. I couldn't run my own life. So on Independence Day, July 4th, 1976—I had no idea ahead of time I was going to do it—I made my decision. I quit Xerox. My last day at the job I actually took my watch off and left it on the desk—I've never put one on since." Back home in Seattle, he started a sales training business and became his own boss.

Rich's involvement with Cell Tech began with an ad in the newspaper and a gift to a friend. "When I was in Seattle after my adventures with the motor home," recalls Loren Spector, "I noticed an ad looking for someone to market training seminars. I attended one of the seminars and met Rich. I liked his work, and we hit it off, so I began selling the seminars for him to large corporations and real estate firms.

"Rich was training from eight in the morning and doing a network marketing thing till ten or 11 at night. I saw the guy was really burning out, so one day I brought a couple of bottles of algae in and said, 'Rich, look. Take this stuff.' And that's how he got started on the product."

Interestingly, there was no thought at the time on either Loren's or Rich's part of Rich doing anything about it as a business. "He was a corporate trainer for another MLM company and had so much visibility that if he'd become a distributor for Cell Tech they would probably have killed him. So it surprised the hell out of me when he showed up at the August Celebration last year, ready to go for it as a business."

How did it happen? In August of 1992, a few years after they'd become regular consumers of the products, Rich and Donia took time out for a vacation down the Oregon coast. They also wanted to see Crater Lake. Having heard something about the August Celebration and realizing they were near Klamath Falls, they decided on the spur of the moment to stop by. They happened in on a Double Diamond panel discussion, and also heard the same John Robbins presentation that Jamie and I had attended. "We were so impressed," Rich recalls, "with the quality of the people we met, the message, and Daryl and Marta's vision, that we made an instant decision to change our lives and invest all our efforts in helping Cell Tech grow."

Rich Hosking's arrival on the Cell Tech scene wasn't the only factor that sparked Cell Tech's dramatic 1993 growth spurt, but it surely played a part. *"Get Rich Quick!"* became the new tongue-in-cheek motto of Distributor Empowerment Teams across North America as they vied with one another to get Rich to come and teach them all he knew. In some cases, Rich's visit meant the doubling of an individual distributor's business almost overnight. *Cell Tech was on the move.*

"We're in it for life," Rich says. "This company has tremendous potential because it's positioned in front of a huge Green Wave. According to the American Advertising Association, 40 percent of consumers will actually pass up a grocery store product unless it indicates some kind of health benefit—low fat, low cholesterol, low sodium, all natural or organic. By the year 2000, they estimate, over 80 percent of the people will be buying that way.

"So what's going to happen to a business that offers a wild-grown, organic, nothing-bad-added-to-it super-food? The truth is, it's going to get carried to tremendous heights by what's happening in the change of consciousness, not only in North America, but throughout the entire world. We're all tired of being poisoned and having our lifespans shortened, with most of our deaths disease-related rather than from natural old age—hence this incredible, giant Green Wave." Rich sees himself as a mechanic who can play a part in helping the company eventually reach the billion-dollar mark.

With the advent of the immensely popular Hosking trainings, Daryl and Marta decided to make available a comprehensive training system to expand and strengthen their growing team for the global work ahead. They called it Relay Team 2000. It currently consists of five members, three trainers plus Daryl and Marta. All will do outreach seminars across the U.S. and Canada—trainings in network marketing, personal growth, and team building.

"Rich's position on the Relay 2000 team is pivotal," says Daryl, "sharing the network marketing opportunity. We cannot be global leaders and global beggars at the same time. Build your business first and then you'll have the resources to contribute to the lives of others."

Personal growth is encouraged by the Master Key seminar offered by Jim Britt. A craggy six-foot-three, fiftyish and buffed, Jim mixes Oklahoma charm and humor with major doses of common sense. He exhorts his audience to let go of limiting patterns and the need for approval and control. (I can relate to that: I'm the kind of person who used to rehearse, as I walked down the subway steps in New York, "Should I say, 'May I have two tokens?' Or would, 'Two tokens, please,' sound snappier?" If *that's* not a need for approval, I don't know what is!) Jim teaches that when we completely let go of the need for outside approval, complete approval is ours from within.

Like Jim's, Brian Biro's team-building seminars are designed to help facilitate the adaptive changes necessary in society to create a sustainable future. Daryl and Marta first met Brian when he was part of the Anthony Robbins training team. (No better team experience than that—we're talking top-of-the-line empowerment here.)

In some of his sessions Brian uses an exercise in board-breaking as a metaphor. Participants are taught to break an inch-thick pine board with their bare hands. "The key," Brian explains, "is the power of the team. First, they have the model of other team members who have already done it. Second, they learn the strategies, the physics of it. The third element is the meaning they decide to attach to what the board represents. When they succeed, they know it was more than just themselves—they know and feel it was their whole team that broke the board."

The Cell Tech team-building seminars demonstrate how our conventional concept of *executive* can be replaced by the concept of *coach* and *teacher*. That means radically changing our usual ego-oriented approach to success. Instead, we have to work together to bring success first to those in our own team, then to our communities, and finally the world. The seminar also shows how individual initiative is vital to the team effort. In essence, what's accomplished is a reversal of the process by which organizations become bureaucracies.

Clearly, this is about more than algae sales. What Cell Tech is building across North America is a fair-sized cadre of people who are clear about what is in their hearts and minds, people who have a strong connection to one another, and immense energy. In addition to the Birkenstock/granola folks, the Cell Tech family includes a community of Hasidic Jews in Brooklyn, strict Christian Creationists, New Agers, and just about everything in between. So the group is unhampered by any narrowness in collective belief that sometimes occurs when a group shares only one belief system or lifestyle.

To me, having been around these people for almost a year, it's just plain good energy. Cell Tech people aren't bigoted, they're not insecure. I can't say it any better than this: They're *partnership* people. I have found it. Partnership. Eureka. Here is a genuine, real, flesh-and-blood prototype partnership group.

What exactly do I mean by "partnership"? I've referred to it a number of times throughout this book—but haven't really explained.

I first came across the concept back in 1989, in a book called *The Chalice and the Blade* by Riane Eisler. Eisler's book impressed me enough to travel to the author's home for what turned out to be a nine-hour interview. The book assembles proof from a number of branches of science to support a startling new social theory—that we humans weren't meant to cause the pain and suffering to one another that we do, and that for hundreds of thousands of years, humans lived in peace and partnership. Thus these past several thousand years of unrelenting bloodshed have actually been atypical.

The partnership way, which Eisler believes is the natural way to live, involves *linking* with other people and with nature. On the other hand, those who believe the only relationship that's any fun is *ranking*—above someone else—are described in her book as the *dominators*. The basic distinction Eisler makes is between a social system in which people honor the chalice of sharing, and one in which they worship the dominating power of the blade.

Past examples of partnership eras include the Minoan civilization of ancient Crete, the movement begun by Jesus of Nazareth, and some aspects of the Renaissance period. On the other hand, the excesses of the Roman Empire, the Spanish Inquisition, and the domination of nature through the Industrial Revolution exemplify the dominator system of civilization. There's some of both partner and dominator in us all, of course; society is healthiest when the partnership side is in control, unhealthy when the dominator mode is uppermost.

For the past 5,000 years or so, the world has been pulled back and forth between the partnership-oriented society that people naturally gravitate toward, and the resultant repression by those who have a vested interest in the status quo. With the advent of technology, the speed of that cause-and-effect seesaw motion has been increasing rapidly, until today we're in a time of almost complete disequilibrium. Dominator agriculture has impoverished our land and thus our food, dominator industry has polluted our skies and poisoned our water, and dominator weapons threaten to destroy all life on the planet. What's happening, according to Eisler, is that the dominator system is self-destructing.

That seems clear enough. Government doesn't work anymore, the infrastructure is crumbling, violence is on the increase, and the number of poor and starving is increasing every day. Yet conversely, in many ways it's also a fairer and more humane society than we've ever seen before. Eisler theorizes that this is because we're engaged in *a climactic battle between the two systems.*

Because of this global chaos, she points out, right now we have a unique opportunity to make the fundamental shift to a partnership society far more easily and more swiftly than we could if the dominator system were more stable. These are the crucial pivotal years. As Daryl often says, borrowing the Dalai Lama's words, "We are the pivotal generation." It seems likely the fate of our civilization will depend upon grass-roots action by partnership-minded groups that are presently in existence, or that will be formed within this decade.

As soon as I grasped this urgent truth, I began writing on the subject, working with the Center for Partnership Studies, and lecturing to study groups around the country. I even began setting my fiction work some 20 years or so into a future which features a more successful partnership-oriented world. *Where there is no vision, the people perish.*

Standing there at the August Celebration picnic a year ago, I felt as if I'd suddenly happened in on a healthy, functioning microcosm of partnership. At the time I wasn't consciously looking for a group that would fit the partnership model, but I guess I despaired that I was finding less and less of it everywhere I went.

That's why the experience of writing this book has been so affirming—and so necessary to me. You already hear all the bad news—this is to let you know some of the *good news,* as Cell Tech and its friends create it. And it isn't fiction, folks. This is happening.

My hope is that other companies will emulate the kind of organization and action that Cell Tech is undertaking.

CHAPTER 14

From Family to Global Family

"THERE'S AN ENORMOUS NEED FOR interaction, for connection among the people of the world, and this is our opportunity," says Daryl Kollman. "You see, we can't go into the wealthiest neighborhood in Nicaragua and make a connection. So what do we do? We make a connection where we can. And the place we can do that is in the poorest neighborhoods of Nicaragua or Los Angeles."

Perhaps the most important thing about Cell Tech's ongoing and growing program in Nicaragua is the person-to-person contact and trust that's being established between the American home team in Klamath Falls, the French-Canadian distributors in Quebec, and the residents and refugees in southwest Nicaragua. Real friendships are growing; understanding has been born. Acupuncturist Marie-Claude, one of the group from Quebec City, now spends several months a year in Nandaime, providing free services to all who need them, sometimes treating as many as 300 people in a day.

I spoke with Gilles Arbour the other day and he told me the barrio children's school performance continues to improve with their regimen of Super Blue Green Algae. This year it looks as though a whopping 98 percent of the children will complete their work, and their average test score is approaching 80 percent, compared with 64 percent pre-algae. That's a huge change!

Another example of how the Cell Tech family is crossing national borders and influencing youth around the world is through the work of 19-year-old Ocean Robbins. The son of Cell Tech's friend and hero John Robbins, Ocean was only 14 when he joined *Youth Ambassadors of America* and attended his first youth summit in Moscow. He and his group met with Raisa Gorbachev, appeared on Soviet TV every night, and were featured in *TIME* magazine.

Ocean is highly articulate and strongly motivated. Growing up on that isolated island off the coast of British Columbia where his parents moved when they fled the Baskin-Robbins fortune, philosophy, and lifestyle, Ocean imagined that the world was a peaceful place. When the family moved back into civilization and he learned about the global nuclear threat, he became very fearful. "I tried not to think about it much and just be a normal kid, but when a plane would fly overhead I'd run outside thinking, Is it a bomb? Or maybe a missile? The fear was with me 24 hours a day."

Then Ocean experienced an important change. "When I was ten, we moved to California and I found out about a play called 'Peace Child.' It was about Soviet and American children making friends and then spreading the cause of peace all over the world. I got into the cast—and went through a profound shift. My fear dissolved because I was *doing* something about what I was concerned about—the nuclear threat.

"At that time I didn't know much about environmental problems. I knew I got a cough every time I went to Los Angeles where we had some relatives, but I didn't know much of the world was breathing air that was causing problems for their lungs. I didn't know the water was unsafe for billions of people on the planet. I'd always had good water. Later, working on those problems at the youth summits, *I began to see the environmental crisis as an opportunity to bring our world together."* So in another way, Ocean is saying what Riane Eisler says: that global crisis can provide both hope and opportunity for us all—hope for a partnership world.

The plan Ocean has come up with to reach out to youth around the world is low-budget, streamlined and highly effective. At the age of 17 he and a friend organized YES!, Youth for Environmental Sanity. Supported in part by Cell Tech, YES! tours high schools in the U.S. and Canada, as well as Australia and New Zealand, reaching more than 150,000 students a year.

Small groups of kids—usually four, mostly high school age—go out on the road, stay at people's homes and speak at school assemblies. Nobody talks for more than two minutes at a time. First the YES! group enumerates the problems: ozone depletion, deforestation, carbon dioxide buildup, pollution, apathy, toxic wastes, and the link between lost jobs and environmental problems. Then solutions come thick and fast. Recycling, voting, starting school ecology clubs, eating lower on the food chain, boycotting and "buycotting"—supporting environmentally appropriate companies.

Next comes an eight-minute game show parody called *"We're In Jeopardy."* Then each teenage performer shares with the audience a personal concern about the global situation. There's also an exercise in team-building which provides vocal support for members of the audience who are doing something positive. "Fred's recycling; let's hear it for Fred!" A pep rally with a surprise ending.

Does it work? Ocean says kids are really starting to think about the environment. "When we first started touring in 1990, about a fifth of the schools we went to had environmental clubs, averaging five to ten members each. Today there are clubs in 95 percent of the schools, with an average of 20 to 30 members."

"What's happening, do you think?" I asked him. "Why is that?"

"Well," Ocean explained, "if you drop a frog in boiling water, it jumps out right away, but if you drop it into lukewarm water that's *gradually* heating up, it doesn't notice the change, and it stays in there until it dies—or until you save it. I think adults are often like the frog in that slowly heating water: Our environmental problems have been growing slowly for so long, it's hard to recognize them. People who've been born into the crisis can feel it more acutely." He paused. "I think the water's getting pretty hot."

Because of the simplicity of Ocean's plan, it seems to be easily duplicable. As Ocean says, "I can see 75 tours going out at the same time all over the country, each of them talking to about 75,000 students a year. That way we'd reach about half the high school students in the U.S. by the 1995-96 school year. I see it happening. I see that shift from watching the world die while trying to deny, to taking responsibility for making a positive difference."

What Ocean said raised a couple of questions in my mind and I'd like to pose them to you now. If more people provided financial support for Ocean's plan, do you think enough kids could be found to visit every high school on Earth? What effect do you suppose that effort alone might have on our future? Don't forget that high school students will be functioning adults in just a couple of years.

Two fundamental beliefs of the Robbins family—that our environmental crisis can bring us together, and that we should think globally and act locally—are validated by some recent environmental successes right here in the Klamath Basin. On the front lines of the current battle between endangered jobs and endangered species, we've been experiencing in-your-face confrontation between the logging/ranching community on one hand and the environmental community on the other. And Cell Tech's ecoordinator has been smack in the middle of the conflict.

Jim Carpenter describes himself as halfway between hippie and redneck. "I relate well to both groups," he says. Slight in build and wiry, sandy-haired, with a face that's open, attentive, curious, Marta's brother radiates patience and calm. No better way to say it than this: he has the face of the kind of brother we'd all like to have.

"When I was in college," Jim says, "I wasn't homespun, I was preppie—or yuppie, anyway. I studied philosophy and art for a while—guess I was trying to comprehend the significance of life on Earth—but how was I going to express it? Well, I thought, it's not going to be on canvas. Then I figured out that I was actually interested in working on the fabric of the landscape itself. That way, if you're ever going to sculpt anything, you've got the whole palette of the biosphere to do it on.

"I like working close to the ground; it's sort of a niche I've been working into in the company. Ecological, environmental, agricultural interface . . . I have a good feeling for it. I'm basically a loner; I could be a great hermit on our ranch out in the Rockies. But I know that through Cell Tech, I can help people in ways I'd never be able to do otherwise. Millions of people are starving around the world. If you ignore that on the evening news, you hardly know about it, but we're facing global anarchy—or worse."

Among the 15 or 20 groups Jim works with is Headwaters. An organization dedicated to saving our watersheds and forests, Headwaters is led by Cell Tech distributor Julie Norman. Her work on behalf of the forest is so highly regarded that she was chosen to be one of the presenters at President Clinton's 1993 Forest Summit; her talk was televised nationally. To honor Julie's work, Daryl Kollman spoke at a Headwaters benefit sponsored by Cell Tech. Rich Hosking came down from Seattle to give a training and there was some media coverage of the events.

Julie's been a consumer of Super Blue Green Algae for a couple of years. She promotes it to her friends but she's not trying to build a business. Instead, her Cell Tech check each month goes to Headwaters. Some people who became interested in the algae business as a result of Daryl's talk realized that by choosing Julie as their sponsor, they could contribute to her cause—without ever having to donate a dime of their own money. Julie's network doubled within a month. That's a win-win situation for everybody—an idea other socially conscious network marketers may want to consider.

"The strength of Cell Tech and the whole project is that it's based on the foundational principles of the future," Jim Carpenter says. "All the companies are going green; that's the buzzword these days. So Cell Tech has the ability to be a leader in a real sense. We're green—in fact, we're Super Blue Green; that's our bread and butter. We can just spread that around the world.

"Things have been changing fast here in the Klamath area ever since the water issue got to be critical. Water, of course, is the original battle, the eternal battle. Water is life. At one of the water meetings a while back, they invited all the different elements—loggers, ranchers, townspeople, tribal members, environmentalists. One of the Indians got up and said, 'Here's water, here's what we think about it.' Then he poured a drink of water for everybody in the audience, and they all had a chance to realize, as they tasted it, just how much pure, fresh water means to them.

"The drought was at its worst back then; the lake had hit rock bottom. But if things hadn't reached this critical phase, people wouldn't be examining it. There wouldn't be two camps. Last year they were talking about not opening up the irrigation system so as to have water for the endangered sucker fish. Then the dialogue started building, as farmers and ranchers who depend on irrigation went from saying, 'Screw the suckers, who needs those cruddy fish anyway?' to recommending that the Bureau of Land Management buy a ranch and convert it back to wetlands. The more the people looked into it, the more they realized that what they'd been doing wasn't sustainable."

"What they were realizing was that domination of nature doesn't work," I interjected.

"They know it will take some water to fill the wetlands, but then it will act as a sponge and the end result will be that they'll have more water." Since my interview with Jim, the BLM has purchased that ranch and the vital wetlands restoration project is underway, spearheaded by those once thought to be among the most entrenched in the old style of agriculture. Instead they're now taking a lead in intelligent resource management. And it's happening fast.

"One of the most important things I always come back to," Jim says, "is the *speed of thought.* The point is, it happens in an instant. It's right now, instantly communicable. It knows no bounds or barriers. How fast can things turn around? They can turn around as fast as we can change our minds. Which is beautiful." He grins.

"Of course Cell Tech needs to do a lot more. Our focus in the market is very narrow at this point, really. We have more than 30,000 distributors, but that's a speck of dust on the planet. They're nearly all white; socially conscious for the most part yes, but what you've got to do is reach a much wider group."

And Cell Tech is starting to do just that. About a thousand miles south of Klamath Falls, a group of people are working on another agricultural project under very different circumstances. The story is best told by Cell Tech distributor Michael Stewart.

"When Cell Tech decided to earmark ten percent of the algae to worthwhile causes, I was more convinced than ever that this was a company I could work with. Its involvement with Chernobyl, Nicaragua, and the Seeing Eye Dog program in Canada touched my heart, but left me with a gnawing question about our kids—especially the inner-city kids who tend to reach for a spray paint can or a gun, and join a gang before they reach for a book. I felt that if these kids had more choices, they could be an asset to themselves and their community.

"Only one problem: How could I help this come about without getting shot? Frankly, 14-year-olds with crack and guns scare the hell out of me. I'd lie awake at night wondering, 'How can I affect our weakest link—how can I help inner-city kids?' 'Cause I'm white and I don't live in South Central.

"Several months ago, at an algae meeting at a local restaurant, one of the gals in my group mentioned a guy named George Singleton, who runs a program called *Hope L.A. Horticulture Corps*. Hope L.A. takes kids who are ready to make positive changes in their lives and teaches them about pride, self-worth, heritage, empowerment and especially how to grow gardens. She said they have a two-acre plot in the hood, a non-truce gang area where drive-by shootings occur almost daily.

"Gardens! That was the end of the meeting for me—that's all I needed to hear. I went from A to Z in my mind; I had a touchdown right there. I said, 'Give me this guy's number.'"

Michael was impressed with George Singleton, an African-American/Mississippi Choctaw whose personal philosophy is a distillation of Rastafarian, Black Muslim, Baptist, and Bahai faiths. Singleton's program is preparing the kids for entry-level jobs in the agriculture and food industries, urban forestry, nurseries, gardens and landscaping services.

"When I first went up there and hung out with these kids," Michael says, "I felt as though I wasn't even worthy to be there with them. Even though I live a fairly clean life, these kids are clear as glass. You look in their eyes, and you look into the soul of God. These guys are deep; it's unreal. They're close to death on a daily basis; ten shots have been fired into the garden at the kids so far. I mean, they have to have truces just to exist. They have to have a truce with the Crips and the Bloods—and that doesn't account for all the subgroups. When I leave, I get swamped by crack salesmen; it's a zoo there. Every time I go up there I cry. These kids needed the algae yesterday."

Michael told me he learned that Singleton had started a program for inner-city, ex-gang kids at Fremont High in Watts. "And I realized we'd come full circle," says Michael, "because Daryl used to teach at Fremont High. So I hammered out a little proposal for Marta to get algae to the kids, and she immediately said yes." Soon shipments were on the way.

"So do the kids eat the algae regularly?" I asked.

"Are you kidding? It's like a sacrament!" He paused. "My dream now," he continued, "is to get the soil amendment [an algae processing by-product being developed by Jim Carpenter] down here to fortify the weak L.A. soil." Michael is pleased with the project so far. "If we can do it in L.A.", he told me, "we can do it anywhere."

Then just before he hung up the phone, Michael Stewart added this postscript. "You know, the interesting thing is, a bunch of these kids are looking to do the Cell Tech business when they finish school."

Chapter 15
One Rub of the Lamp

A YEAR HAD PASSED. IT was the eve of the August Celebration, and we were getting ready for the week-long party. By now, Tom and I had acquired a large circle of new friends, and over a thousand of them were coming to town. So Tom was painting the house in preparation and we were picking vegetables from our algae-fertilized garden to feed at least a few of them when they got here. In the spring we'd gotten some frozen Cell Tech soil amendment in big drums, thawed it out, and spread it over the land. The grateful soil, depleted no more, offered up a bounty of beets, corn, squash, pumpkins, spinach, lettuce, peas, peppers, and cucumbers, along with bright red tomatoes almost too big to get my two hands around. I walked out in the sun and looked over that bumper tomato crop, thinking about those kids in L.A., wanting everyone to have this kind of bounty.

All the people I've been writing about are going to be in town, I thought—Showshawme, Rich Hosking and Donia, Gilles Arbour and his friend/sponsor Art Robbins. John Robbins (no relation) will be back, as well as his son Ocean and the YES! tour. Afterward, John will be leading distributors on a week-long raft trip down the wild and scenic Klamath River.

Kim Bright and Ray Cassano are coming in from Florida. Ray not only achieved the rank of Double Diamond in record time as Kim had predicted, he then *doubled* the Double Diamond requirements just one month later.

Showshawme's friend from all those years ago, nutritionist Viktoras Kulvinskas, will give a talk, and Father Santiago will be here from Nicaragua to share the progress on the projects there. Explorer Bernard Voyer is planning to show a film of his trip across Ellesmere Island at the North Pole on skis, and Yves LaForest will tell us about his successful Everest climb—both feats accomplished with Super Blue Green Algae.

It's been quite a year since I happened upon the last Celebration. As I prepare to wrap up this book, I'm struck by the many different things that the algae means to different people—from the frivolous to the profound. Just one rub of the lamp seems to bring so much.

For the athletes, algae provides a way to win the race. For Julie Norman it's a way to help protect the forests she loves; for Loren Spector, a chance to play all the tennis he wants. For a bunch of college students it's a way to pull through those tough study sessions. For our friend Joe Yellowhawk, Super Blue Green has meant a way to lose 100 pounds without even trying. For Rich Hosking, it represents that trip to Bora Bora he's always wanted to take. For Showshawme, it's having a family at last. For Gilles Arbour, a way to save the children. For most everyone I've met, it's a food, a business, and a cause all rolled into one.

In my own family, the algae seems to have opened a new chapter. Tom has entered an expanded world of artwork and creativity. Jamie's reading speed has improved. Chronic fatigue interferes less with Steven's work schedule, and my daughter Cindy's mood swings seem to be mostly a thing of the past. No way to know for sure whether algae has made the difference. We can't help but believe it's played a role.

For me, algae has brought clarity, better health, more productive hours in every day—and most important, it's brought me *hope.* After years of striving without much visible effect, I now feel part of a partnership network that can have an influence, perhaps a strong influence, for a better future. For after all, these days "power" usually has the word "corporate" in front of it. It's business, not government, that's running things. And it's business, one way or another, that will see us into the 21st century.

I was talking with Marta about that the other night. We were sitting on the stairs just outside her office on the bridge. It was after hours, about 7:30 PM, the best time for that wonderful old building. The colors, the columns, the silent computers, and the deserted shipping department out beyond... you could almost sense it resting from another day with an unprecedented number of orders. Lately, the company's been growing at about ten percent per month. One of these days soon, I'm sure, instead of quietly nestling into a number of buildings around town, Cell Tech will be headquartered in one huge state-of-the-art plant in a spectacular park of its own. Planted with vegetables and sprouts, I would bet, if I know Daryl and Marta.

"It's too bad, in one way. This is so homey," I told Marta. She agreed. Marta enjoys her time alone in the office evenings and weekends, running thousands of commissions through the printer, catching up on administration, and listening to her endless voice mail.

We sat there for a few minutes, talking about tomatoes, and about Tom's latest antler sculpture (Marta is a great fan of Tom's—she owns his first set of kachina dolls). "One day a few weeks ago," I told her, "he started carving faces that were singing—and now there are 12 of them along the curve of the antler. Native Americans, young, old, in different tribal dress. The singers seem to be telling us to take care of our land—one of them's offering a peace pipe."

"What does he call it?"

"*Ancestral Voices.*" We were both quiet for a moment. "So what do *you* see for the future?" I asked her then.

Like most mothers, she automatically thought of her kids and what she wants for them—those three little boys who learned to count to 100 by packing bottles of capsules at the kitchen table. "After all," she said, "they were the ones whose lives were influenced the most during all those years in New Mexico with the foam rubber for a sofa and one pair of shoes each."

Of Daryl and Marta's four kids, at least two will be involved long term with the running of Cell Tech. Joe, who heads production, tends to look at things from a business point of view—he sees an almost limitless business opportunity in the algae pond project. The plan, Joe once explained to me, is for farmers all over the world to maintain small algae ponds in every field, using the algae as a food source and plowing the remainder into the earth to enhance the nutritional value of their crops. Thus we could reduce our dangerous carbon dioxide levels while revitalizing our soil and improving the nutritional quality of our food.

Then there's Justin, about to graduate from the University of Arizona with a degree in communications. Just as his mother always knew the family would buy Cell Tech back, Justin has always known he would make a career of Cell Tech. He seems suited to it—he's outgoing, personable—and like his mother often quite funny. (Marta's sense of timing and delivery can be devastating.) Peter, Daryl's older son, is a physics whiz at Cal Poly—and with genes like Daryl's, who knows what kind of projects he'll come up with? Sevin, Marta's oldest, whose specialties are computer science and Spanish, has his own consulting business in Chicago. So it looks like among their four talented progeny, Daryl and Marta have all bases covered.

"But what about *you?*" I persisted. "What do *you* want to work on?" I had to phrase it that way, for whatever Marta does with her life, it will always be work—work is her fun. She thought for a moment, considering my question.

"One thing that bothers me is that we seem to be giving up our freedoms and don't even realize it. Right now," she pointed out, "the Food and Drug Administration seems to be trying to regulate even what herbs we use. If we don't watch it, we're going to wake up some day to discover we need a prescription for Vitamin C. Or that chamomile tea has become a controlled substance."

I agreed. "It's called others trying to dominate our life."

Marta continued. "We absolutely have to take responsibility for ourselves, for our choices, for our children's choices. And for their ability and right to make those choices. But we're standing silently by, watching our educational system rot."

She's right of course. And in my view, it's the ignorance born of our growing illiteracy that's destroying democracy. An uneducated people can't govern themselves; we see that on the news every day.

Marta continued. "We send our most valuable resource to spend most of the day in a system that needs major attention. Many of our children can't read, can't think for themselves. They aren't being taught how to solve problems. We're all letting our children down—the parents, the schools, and the society that we expect them somehow to move smoothly into as young adults."

Then Marta was onto a related subject, talking about partnership in the most basic of our relationships, male and female. "I don't think we have to be anti-male. It concerns me. We don't have to be against men to elevate ourselves. It's an individual thing. To me it isn't a group of women getting together so much as it's every woman sitting down and saying, 'Who am I? What's my purpose? What do I want to accomplish?' And realizing she is the only one who can do that. Then doing something about it. Peacefully, gently, but persistently. Every single solitary day." Of course that's how Marta became successful. And more success was about to come.

During the August Celebration a few nights later, Marta received a surprise, the first "Greatest Networker" award from the network marketing magazine, *Upline*. "Of all Marta

Kollman's achievements," the citation read, "the most profound has been her recognition and development of the wealth of talent within the distributor network. Under her direction, the Cell Tech corporate staff and leadership embodies the concept of *servant-leader*."

John Milton Fogg, publisher of *Upline*, also pointed out that Cell Tech is 100 percent on-trend, according to guidelines set out by author and corporate consultant Faith Popcorn, the woman who's been 95 percent correct in predicting trends in America over the past 20 years. Out of the ten trends that she describes, nine might have been written to describe Cell Tech. I was heartened. Unlimited product, unlimited market, and perfect timing—what more could you want?

In Daryl's Saturday evening program, always one of the highlights of the Celebration, Father Santiago provided a surprise for the distributors. Though none of their Cell Tech friends were in Nicaragua, the people of Nandaime had thrown a street party for all of us, complete with entertainment, and had made us an amateur video of the event. Even the poor lighting conditions couldn't hide the beauty of the festivities or the sincerity with which they were shared.

Later Daryl spoke about his priorities for the years ahead. "Over the past 2,000 years," he said, "it took 1,900 years for the world population to double, from 600 million to 1.2 billion." He began to draw the curve on a blackboard. "It took only 55 years for it to double again, and then just 30 years to double once more. That's a rate of growth that is *not sustainable*," he said, pointing to the top of the curve, which was headed nearly straight up. "This is one of the reasons that we're the pivotal generation.

"And when you look at carbon dioxide concentration, it was at 290 parts per million 10,000 years ago. That figure was constant up to 1900, then it went from 290 to 295, then to 320, then 350. That's not sustainable, either. There are 200 billion extra tons of carbon dioxide between sea level and 15,000 feet, and it's increasing so rapidly that it's drastically changing our weather." Daryl then talked about the unprecedented floods on the Mississippi that were still wreaking havoc as he spoke.

"Isn't it interesting," he went on, speaking to a full house in the darkened theater, "that we're faced with these two challenges. A population growth that's going through the roof and a carbon dioxide concentration that's doing the same thing. These, in my opinion, are the two main issues facing the pivotal generation.

"There's also an increasing disparity between the very rich and the very poor. In 1940, ten percent of the people controlled 40 percent of the wealth in this country. In 1990, *one* percent of the people controlled 40 percent of the wealth. We have an enormous number of poor people on the planet, and that's growing on the same curve as population and carbon dioxide. It's going to take incredible wealth to turn things around—it will take 500,000 square miles of algae ponds just to start reducing the carbon dioxide. We have the wisdom and the technology; we just have to choose to do something."

Daryl continued. "To extend the window of opportunity so that our children have one as well, we must make a commitment, here, tonight, to double the size of our company this coming year. If we do that we'll grow to 60,000 distributors by next year. The second year we'll grow to 120,000 distributors, the third year to 240,000. Not an insignificant group when we have a quarter of a million people in Cell Tech. We'll have a lot more money to do a lot more things." He paused and looked at the audience. "The only reason I'm talking about money [Daryl rarely talks about money] is because more than 50 percent of it will be back in your pockets, and you're going to be able to contribute to the lives of others with that money." Loud applause burst out.

As it died down, Daryl said, "This is a very powerful, very organized, very contributing group. We're different from so many other groups. We aren't *against* things, we are *for* things, and that makes a real difference. So I want to ask you tonight, are you willing to make that commitment? Are you willing to commit to doubling every year for the next ten years?" Applause, shouting, foot stamping, and wild cheering. "Yes!" shouted Daryl, above the crowd. "My 27 trillion cells are celebrating!

"We can't do this alone. This is a large planet. It's utterly and absolutely impossible to change the length of the window of opportunity alone. A single country cannot do it. A single person cannot do it. The YES! tour is right; what's happening in our environment will bring us together probably for the first time ever. We're going to experience this thing called peace, and guess what? We're going to like it—we're going to love it!

"We're going to switch, in a single generation, from thinking in terms of competition to thinking only in terms of cooperation. We're going to have a generation growing up beyond our children's generation that won't have a concept or a feeling about war," he said, holding his arms out to the audience. "They just won't know what it is that we're talking about. Won't that be incredible?" The people were on their feet, cheering.

The Celebration closed with the traditional candlelight ceremony. Everyone holds a candle, and one is lit. That person lights another candle, those two people light two others, and so on. "Get the concept of how slowly things move at first," Daryl remarked to the audience, as Marta lit his candle, "and how rapidly they move after a few seconds." Within moments, about a thousand candles illuminated the entire auditorium. It was breathtaking.

On the way home, I asked Tom, "Did you notice how quickly that light spread once it got started?"

He nodded. "It's the power of doubling," he replied. "Do you remember the story about the beggar who was rewarded by the king?"

"No," I said.

"The beggar said, 'Your majesty, all I want is one grain of rice on the first square of this chessboard,' (the king was playing chess at the time) 'two grains on the next square, four on the next, eight on the next . . .'—and so on, doubling through all 64 squares."

"So what happened?"

"It turned out to be enough rice to cover the whole kingdom three feet deep."

"Wow. *Doubling.* So that's what Daryl and Marta have in mind."

The day the the Celebration was over, I decided to go for a run to think it all out. I put on shorts and a T-shirt, grabbed my trusty pocket dictaphone in case I thought of something deathless along the way, and set out toward the lake. It was a beautiful day—the sky was clear, and there was just enough of a breeze to make the run pleasant.

Is it naive, I asked myself, as I made my way up the little rise on Third just before California Street, to believe that algae can help save the world? Doubtless some will say it is. But nothing else has proven to do any good. As Daryl says, "There was never a bullet designed that could extend our window of opportunity one second." Bullets can't save our world, and all the people who put their faith behind force are tragically mistaken. The violence in every aspect of our society is spreading like wildfire—so much so that many of us fear for our survival. Yet as John Robbins, Riane Eisler and Daryl Kollman all have pointed out, times of crisis are also fertile times for change.

I thought about New York City, where I lived for so many years, and what it used to be like during blackouts, blizzards and those other disasters New Yorkers are so fond of telling stories about. At these times, it always seemed that everybody in the city, almost without exception, was bent on helping someone else. I am so certain that we all *want* to operate that way with one another. So why don't we? I heard again in my mind Rodney King's plaintive words, "Can't we all just get along?"

I was running over the Link River Bridge now, looking down into the water below as countless millions of strands of algae flowed past me in a swift-moving stream of bright rich green. The smell of it here close to the water was fresh and clean, so everyday-familiar to me now; it's become part of every aspect of my life. And to think that a year ago, it wasn't even in my consciousness. It's as Bertrand Russell said, "*The universe is full of magical things just waiting for our wits to grow sharper.*"

Over this past century, we've done terrible damage both to our planet and to our health; it's clear to most of us that we need to clean up the environment to save our children and ourselves. But it may be just as important to improve our internal environment. People dulled by nutritional deficiency often become sluggish, hostile, selfish—or violent. They tend to want more and more comforts, and if they can afford it, they buy more recliners, turn up their thermostats—and turn down their sensitivity to others' suffering. On the other hand, those whose minds and bodies are clear tend to eat lower on the food chain, have more energy, live lighter on the planet, and have more love to spare.

I have to believe that there's still time to turn things around—and that algae can help. If algae ponds can convert excess carbon dioxide to oxygen about a jillion times faster than trees and plants, then why aren't we doing it!? And if eating algae can make us healthier, can help us re-connect with one another and with the Earth, then let's try it. After all, this isn't just some new product or vitamin supplement. It's Earth's *first food* that we're talking about. That we'd *forgotten* about until Daryl and Marta rediscovered it. We need to make use of that molecule of hope.

Entering Moore Park where the Celebration picnic was held, I jogged up through the avenue of stately trees that were planted in the early part of the century. As I reached the crest of the hill, I looked out over the mirror-calm lake and the snow-capped mountains beyond. Past my halfway mark now and starting toward home, I thought about Kanoa

Christopher Kelii, my brand new dark-haired, bright-eyed grandchild. I haven't seen him yet, but I will, at Thanksgiving. And one of the first things I'm going to do—I'm going to bring him down to this magical lake, and show it to him.

* * *

About the Author

LINDA GROVER is the author of two other books. *The House Keepers* is a humorous tale about the residents of her Manhattan apartment building who fought City Hall to save their home, and won. It was published by Harper and Row, and serialized in the *New York Post*. *Looking Terrific* (with Emily Cho), published by Putnam's and Ballantine, is a book about the language of clothing. It was a Literary Guild selection and a *New York Times* national best-seller. Linda has also been a head writer of television drama for NBC, CBS, and ABC. She and her partner, artist Tom Daugherty, make their home on a hill in Klamath Falls, Oregon.